Applied Probability

AMS/IP

Studies in
Advanced
Mathematics

Volume 26

Applied Probability

Proceedings of an IMS Workshop
on Applied Probability
May 31, 1999–June 12, 1999
Institute of Mathematical Sciences at the
Chinese University of Hong Kong
Hong Kong, China

Raymond Chan, Yue-Kuen Kwok,
David Yao, and Qiang Zhang,
Editors

American Mathematical Society · International Press

Shing-Tung Yau, General Editor

2000 *Mathematics Subject Classification.* Primary 46N30, 47N30, 28Dxx, 93Exx, 91Bxx.

Library of Congress Cataloging-in-Publication Data

IMS Workshop on Applied Probability (1999: Chinese University of Hong Kong)
Applied probability : proceedings of an IMS Workshop on Applied Probability, May 31, 1999–
June 12, 1999, Institute of Mathematical Sciences at the Chinese University of Hong Kong, Hong
Kong, China / Raymond Chan. . . [et al., editors].
    p. cm. — (AMS/IP studies in advanced mathematics, ISSN 1089-3288 ; v. 26)
  Includes bibliographical references.
  ISBN 0-8218-3191-7 (alk. paper)
  1. Probabilities—Congresses. I. Chan, Raymond, 1958– II. Title. III. Series.

QA273.A1I48   1999
519.2—dc21
                                                              2002025578

---

10 9 8 7 6 5 4 3 2 1   07 06 05 04 03 02

Contents

A Direct Method for Stochastic Automata Networks
Raymond Chan and Wai Ki Ching 1

Estimating the Speed of Random Walks
Dayue Chen 17

A New Story of Ergodic Theory
Mu-Fa Chen 25

Solvability of a Stochastic Linear Quadratic Optimal Control Problem
Shuping Chen and Jiongmin Yong 35

Convertible Bonds with Market Risk and Credit Risk
Mark Davis and Fabian R. Lischka 45

Quasi-Monte Carlo Methods and Their Randomizations
Fred Hickernell and Regina H.S. Hong 59

Contingent Claim Approach for Analyzing the Credit Risk of Defaultable Currency Swaps
Hong Yu and Yue-Kuen Kwok 79

Dynamic Insider Trading
Shunlong Luo and Qiang Zhang 93

A New Hedging Model and a Nonlinear Generalization of Black-Scholes Formula
Shanjian Tang 105

An Overview on Martingale Approach to Option Pricing
Jia-an Yan 121

On Comparison Theorems for Diffusion Processes
Xinsheng Zhang 135

Foreword

The *"IMS Workshop on Applied Probability"*, organized by the Institute of Mathematical Sciences at the Chinese University of Hong Kong, took place in Hong Kong on May 31 to June 12, 1999. The aim of this two-week workshop was to promote research interest in applied probability for local mathematicians and engineers and to foster contacts and exchanges with experts from other parts of the world. The main themes of the workshop were: Mathematical Finance (first week) and Stochastic Networks (second week).

This book contains eleven papers from the invited speakers of the workshop. The topics range from the theoretical study such as ergodic theory and diffusion processes to the very practical problems such as convertible bonds with market risk and insider trading.

The editors of the book are Raymond Chan (The Chinese University of Hong Kong), Yue-Kuen Kwok (Hong Kong University of Science and Technology), David Yao (The Chinese University of Hong Kong), and Qiang Zhang (City University of Hong Kong). On behalf of the organizers of the workshop, we would like to thank the following sponsors for their generous support: The Institute of Mathematical Sciences and the Department of Systems Engineering and Engineering Management at the Chinese University of Hong Kong, and also The Zheng Ge Ru Foundation.

AMS/IP Studies in Advanced Mathematics
Volume 26, 2002

A Direct Method for Stochastic Automata Networks

Raymond H. Chan and Wai Ki Ching

ABSTRACT. Stochastic Automata Networks (SANs) are widely used in modeling communication systems, manufacturing systems and computer systems. The SAN approach gives a more compact and efficient representation of the network when compared to the stochastic Petri nets approach. To find the steady state distribution of SANs, it requires solutions of linear systems involving the generator matrices of the SANs. Very often, classical direct methods such as the LU decomposition are inefficient because of the huge size of the generator matrices. An efficient algorithm should make use of the structure of the matrices. In this paper, we present a direct method based on taking the circulant approximation of the generator of SAN.

1. Introduction

Stochastic Automata Networks (SANs) are widely used in modeling queueing systems [4, 5, 6, 28], communication systems [27, 29, 35], manufacturing systems and inventory control [7, 20, 21, 22, 23], computer networks [31] and of course discrete-event systems [2, 25]. The SAN approach has a more compact and efficient representation when compared to the stochastic Petri nets approach [1, 31]. Moreover because of the special structure of the resulting representations, matrix-vector multiplications involving the generator matrices can usually be done very fast [34].

In analyzing the system performance of a SAN, it is required to find its steady state distribution, which can be obtained by solving a linear system involving the generator matrix of the SAN. In general the solution cannot be obtained efficiently by classical direct methods such as the LU decomposition due to the huge size of the generator matrix. Efficient numerical algorithms should make use of the special structures of the generator matrices and their fast matrix-vector multiplications. Iterative methods based on circulant preconditioning techniques were studied in [12, 19]. In this paper, we present a direct method for solving the steady state probability distribution of SAN. The direct method is based on taking circulant approximation of the generator matrix of SAN and the Sherman-Morrison formula, [26, p.3].

Circulant matrices are particular case of Toeplitz matices. Toeplitz matrices are matrices with constant diagonal entries. Circulant matrices are Toeplitz matrices such that each column is a cyclic shift of its preceding column. One important

Key words and phrases. Stochastic Automata Networks, Circulant approximations.

property of circulant matrices is that they can be diagonalized by Fast Fourier Transforms [24]. Hence their inverses can be found easily. Circulant matrices have shown to be good preconditioners for Toeplitz systems in many applications [13] and in particular in queueing networks, see for instance [11, 14, 15, 16, 20]. The main observation in queueing network applications is that most queueing networks have generator matrices that are close to Toeplitz matrices. These include sophisticated networks such as the Markov modulated Poisson processes arising in manufacturing systems and inventory control systems and also networks with more general queueing disciplines such as batch arrivals.

The paper is organized as follows. In §2, we first introduce a two-queue overflow network which is a particular example of SANs. In §3, we give an introduction of SANs. We then construct our circulant approximations for SANs in §4. In §5, we discuss our methods for three practical examples of SANs. Finally, concluding remarks are given in §6.

2. An Overflow Queueing Network

Let us begin with some notations. We will use **0** and **1** to denote the zero column vector and the column vector of all ones of appropriate length respectively. Also we will use O and I to denote the zero and identity matrices of appropriate size respectively. For any matrix A, $z(A)$ will denote the number of nonzero columns in A. A matrix A is said to be nonnegative, denoted by $A \geq O$, if all the entries of A are nonnegative.

To introduce the terminologies and notations of SANs, let us consider a simple example of SANs with 2 automata. It is the 2-queue overflow network considered in [8, 28]. The network consists of two queues (*automata*) with exogenous Poisson arrivals and exponential servers. Whenever queue 2 is full, the arriving customers will overflow to queue 1 if it is not yet full. Otherwise the customers will be blocked and lost, see Figure 1.

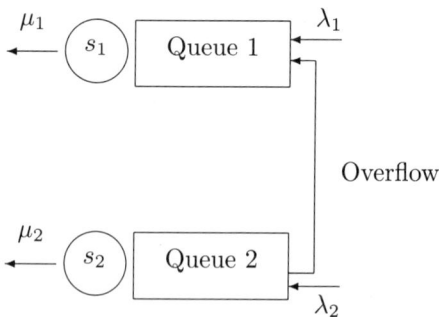

Figure 1. The Two Queue Overflow System.

For queue i, $i = 1, 2$, let λ_i be the exogenous input rate, μ_i the service rate, s_i the number of servers, and $l_i - s_i - 1$ the buffer sizes. Then the generator matrix for the queueing system is given by

(1) $$A = Q_1 \otimes I_{l_2} + I_{l_1} \otimes Q_2 + R \otimes \text{Diag}(0, \cdots, 0, 1),$$

where

(2)

$$
Q_i = \begin{pmatrix}
\lambda_i & -\mu_i & & & & & & 0 \\
-\lambda_i & \lambda_i + \mu_i & -2\mu_i & & & & & \\
& \ddots & \ddots & \ddots & & & & \\
& & -\lambda_i & \lambda_i + s_i\mu_i & -s_i\mu_i & & & \\
& & & \ddots & \ddots & \ddots & & \\
& & & & -\lambda_i & \lambda_i + s_i\mu_i & -s_i\mu_i \\
0 & & & & & -\lambda_i & s_i\mu_i
\end{pmatrix}, i = 1, 2,
$$

(3)

$$
R = \begin{pmatrix}
\lambda_2 & & & & 0 \\
-\lambda_2 & \lambda_2 & & & \\
& -\lambda_2 & \ddots & & \\
& & \ddots & \lambda_2 & \\
0 & & & -\lambda_2 & 0
\end{pmatrix}
$$

and I_{l_i} is the identity matrix of size l_i, see [28] for instance.

We note that the matrix Q_i in (2) corresponds to the generator matrix of a stand-alone queue i and hence the matrix $(Q_1 \otimes I_{l_2} + I_{l_1} \otimes Q_2)$ in (1) corresponds to a 2-queue network where no overflow can occur. It is called the *non-interlacing* part of the network. The last term $R \otimes \mathrm{Diag}(0, \cdots, 0, 1)$ in (1) corresponds to where the overflows (or *transitions*) occur. In general, SANs are composed of the non-interlacing part of the network together with the transitions allowed. For our 2-queue overflow network, the queueing disciplines are governed by three probabilistic rules, namely, the Markovian input-output processes of queues 1 and 2 (the non-interlacing part) and the overflow process from queue 2 to queue 1. In fact, the generator matrix A in (1) can be written in the form

$$
A = \sum_{i=1}^{3} \bigotimes_{j=1}^{2} Q_{ij}
$$

where $Q_{11}, Q_{21}, Q_{31}, Q_{12}, Q_{22}$ and Q_{32} are $Q_1, I_{l_1}, R, I_{l_2}, Q_2$ and $\mathrm{Diag}(0, 0 \cdots, 0, 1)$ respectively. Here for ease of presentation, we denote

$$
\bigotimes_{j=1}^{k} H_i = H_1 \otimes H_2 \otimes \cdots \otimes H_k.
$$

We note that A can be rewritten in a more complicated form, which is however the standard form of SANs:

(4)
$$
A = \sum_{i=1}^{3} \left\{ \bigotimes_{j=1}^{2} D_{ij} - \bigotimes_{j=1}^{2} E_{ij} \right\},
$$

where

$$(5) \quad D_{ii} = \begin{pmatrix} \lambda_i & & & & & & 0 \\ & \lambda_i + \mu_i & & & & & \\ & & \ddots & & & & \\ & & & \lambda_i + s_i\mu_i & & & \\ & & & & \ddots & & \\ & & & & & \lambda_i + s_i\mu_i & \\ 0 & & & & & & s_i\mu_i \end{pmatrix}, \quad i = 1, 2,$$

$$(6) \quad E_{ii} = \begin{pmatrix} 0 & \mu_i & & & & 0 \\ \lambda_i & 0 & 2\mu_i & & & \\ & \ddots & \ddots & \ddots & & \\ & & \lambda_i & 0 & s_i\mu_i & \\ & & & \ddots & \ddots & \ddots \\ & & & \lambda_i & 0 & s_i\mu_i \\ 0 & & & & \lambda_i & 0 \end{pmatrix}, \quad i = 1, 2,$$

$$(7) \quad D_{31} = \begin{pmatrix} \lambda_2 & & & 0 \\ & \lambda_2 & & \\ & & \ddots & \\ & & & \lambda_2 \\ 0 & & & 0 \end{pmatrix}, \quad E_{31} = \begin{pmatrix} 0 & & & 0 \\ \lambda_2 & 0 & & \\ & \lambda_2 & \ddots & \\ & & \ddots & 0 \\ 0 & & \lambda_2 & 0 \end{pmatrix},$$

$$(8) \quad D_{12} = E_{12} = I_{l_2}, \quad D_{21} = E_{21} = I_{l_1}, \quad \text{and} \quad D_{32} = E_{32} = \mathrm{Diag}(0, \cdots, 0, 1).$$

We see that in this standard form, D_{ij} and $E_{ij} \geq O$, and that D_{ij} are diagonal matrices with diagonal entries equal to the column sums of E_{ij}. In particular, D_{ij} and E_{ij} have the same column sums, i.e. $\mathbf{1}^t D_{ij} = \mathbf{1}^t E_{ij}$ for $i = 1, 2, 3$, $j = 1, 2$.

To analyze the 2-queue network, we need to find its steady state distribution vector, which is the normalized right null vector of A. More precisely, the vector $\mathbf{p}$ is the nonnegative vector that satisfies $A\mathbf{p} = \mathbf{0}$ and $\mathbf{1}^t\mathbf{p} = 1$. Classical methods such as the block Gauss-Seidel method and the successive over-relaxation method are standard methods for solving this problem, see [28]. In [8, 14], the preconditioned conjugate gradient method is used and a very efficient preconditioner is constructed for networks where the traffic is about balance, i.e. $\lambda_i \approx s_i\mu_i$.

3. Stochastic Automata Networks

A Stochastic Automata Network (SAN) consists of a number of individual stochastic automata. Each automaton is represented by a number of states and probabilistic rules that govern the transitions from one state to another. The state of an automaton at time t is just the state it occupies at time t and the state of the SAN at time t is given by the states of its constituent automata. For more details of SANs, see [4, 5, 6] for instance.

Consider a SAN with n automata and m probabilistic rules. Let the state space of the ith automaton be of size $l_i - 1$. The generator matrix A of the SAN can be written in the standard form:

$$(9) \qquad A = \sum_{i=1}^{m} \left\{ \bigotimes_{j=1}^{n} D_{ij} - \bigotimes_{j=1}^{n} E_{ij} \right\},$$

where D_{ij} and E_{ij} are of size l_j-by-l_j, see [**4, 31, 34**] and cf. (4). Here $E_{ij} \geq O$ are such that each term $\bigotimes_{j=1}^{n} E_{ij}$ represents the rate of certain transitions amongst the states. The D_{ij} are nonnegative diagonal matrices that contain the column sums of E_{ij}, i.e.

$$(10) \qquad \mathbf{1}^t D_{ij} = \mathbf{1}^t E_{ij}, \quad i = 1, \dots, m, \ j = 1, \dots n.$$

For simplicity, let the non-interlacing part of the network be represented by the first n terms in (9). Thus, $D_{ii} - E_{ii}$ is the generator matrix of the ith automaton alone and $D_{ij} = E_{ij} = I$, for $1 \leq i \neq j \leq n$ (cf. the first two terms in (1)). More precisely, the first n terms of (9) can be written as

$$(11) \qquad (D_{11} - E_{11}) \otimes I \otimes \cdots \otimes I + I \otimes (D_{22} - E_{22}) \otimes I \otimes \cdots \otimes I + \cdots$$
$$(12) \qquad + I \otimes \cdots \otimes I \otimes (D_{nn} - E_{nn}).$$

For ease of discussion, we assume that the generator matrices $(D_{ii} - E_{ii})$, $i = 1, \dots, n$, of the individual automata are irreducible. As D_{ii} are diagonal matrices, E_{ii} are therefore irreducible. Clearly from (11), we see that $\sum_{i=1}^{n} \bigotimes_{j=1}^{n} E_{ij}$ is irreducible. Using the fact that $E_{ij} \geq O$, we see that $\sum_{i=1}^{m} \bigotimes_{j=1}^{n} E_{ij}$ and therefore A in (9) are irreducible too. We thus have proved the following proposition.

PROPOSITION 1. *If the generator matrix of each of the individual automata in a SAN is irreducible then the generator matrix of the SAN (i.e. A in (9)) is also irreducible.*

To analyze the network, we need to find its steady state distribution vector $\mathbf{p}$, which is the normalized right null vector of A. We note that if $n = m$, i.e. the network consists only of the non-interlacing part (11), then $\mathbf{p}$ can be obtained easily by taking the tensor product of the steady state distribution vectors of the individual automata. For the general case where $m > n$, the existence of $\mathbf{p}$ follows from the irreducibility of the matrix A, as the following proposition shows.

PROPOSITION 2. *Let B be an irreducible matrix of the form*

$$B = \sum_{i=1}^{m} \left\{ \bigotimes_{j=1}^{n} U_{ij} - \bigotimes_{j=1}^{n} V_{ij} \right\},$$

where $V_{ij} \geq O$ and U_{ij} are diagonal matrices such that

$$(13) \qquad \mathbf{1}^t U_{ij} = \mathbf{1}^t V_{ij}, \quad i = 1, \dots, m, \ j = 1, \dots n.$$

Then B has a 1-dimensional null space with positive null vectors.

PROOF. It follows from (13) that $\mathbf{1}^t(\bigotimes_{j=1}^{n} U_{ij}) = \mathbf{1}^t(\bigotimes_{j=1}^{n} V_{ij})$ and hence $\mathbf{1}^t B = \mathbf{0}^t$, i.e. B has zero column sums. Clearly B has nonpositive off-diagonal entries. In particular, the column sum of the off-diagonal entries of any column of B cannot be positive. Since B is irreducible, these column sums cannot be zero either. Because B has zero column sums, the diagonal entries of B are therefore positive. Let D be the diagonal matrix containing the diagonal entries of B. Then

$I - BD^{-1}$ is an irreducible column stochastic matrix. The proposition now follows from the Perron-Frobenius theorem, see for instance [**3**, p.27]. $\qquad\qquad\square$

By Proposition 2, we see that the steady state distribution vector $\mathbf{p}$ for A in (9) exists and is unique. Moreover, all the entries in $\mathbf{p}$ are positive. Classical iterative methods such as the block Gauss-Seidel method and the successive over-relaxation method are standard methods for finding $\mathbf{p}$ [**28, 29, 34**]. However, in this paper, we consider a direct method based on taking circulant approximation of the generators.

We note that because $\mathbf{p}$ exists and is unique and positive, it can be obtained by normalizing the solution $\mathbf{x}$ in the matrix equation

$$(14) \qquad\qquad G\mathbf{x} \equiv (A + \mathbf{e}\mathbf{e}^t)\mathbf{x} = \mathbf{e},$$

where $\mathbf{e} = (0, 0, \cdots, 0, 1)^t$, i.e. $\mathbf{p} = \mathbf{x}/(\mathbf{1}^t\mathbf{x})$. In the following, we will consider approximations for G. Clearly, we have

$$(15) \qquad\qquad \mathrm{rank}(G - A) = 1.$$

4. Circulant Approximation for SAN

Toeplitz matrices are matrices with constant diagonal entries. Circulant matrices are Toeplitz matrices such that each column is a cyclic shift of its preceding column. One important property of circulant matrices is that they can be diagonalized by Fast Fourier Transforms [**24**]. Hence their inverses can be found easily. Circulant approximations have been used in many applications where the Toeplitz matrices come into play, such as in image processing, partial differential equations, integral equations and in particular queueing networks, see [**13**] and the references therein. In this section, we consider the construction of circulant approximation for SANs.

The success of our approximation depends on the observation that in many network applications, the matrices D_{ij} and E_{ij} in (9) are low rank perturbations of Toeplitz matrices (cf. (5)–(8)). Hence they can be approximated well by circulant matrices. Our idea of constructing the approximations is to solve the first automaton and its related disciplines approximately and the remaining automata exactly. More precisely, we will approximate the matrices E_{i1} in (9) by nonnegative circulant matrices $c(E_{i1})$ that are low rank perturbations of E_{i1}.

For example, if E_{i1} takes the form of E_{ii} in (6), its circulant approximation will be given by

$$(16) \qquad c(E_{ii}) = \begin{pmatrix} 0 & s_i\mu_i & & & & & \lambda_i \\ \lambda_i & 0 & s_i\mu_i & & 0 & & \\ & \ddots & \ddots & \ddots & & & \\ & & \lambda_i & 0 & s_i\mu_i & & \\ & 0 & & \ddots & \ddots & \ddots & \\ & & & & \lambda_i & 0 & s_i\mu_i \\ s_i\mu_i & & & & & \lambda_i & 0 \end{pmatrix}.$$

It is a rank s perturbation of E_{ii}. In fact, the number of nonzero columns of $(c(E_{ii}) - E_{ii})$, denoted by $z(c(E_{ii}) - E_{ii})$, is equal to $s+1$. For E_{31} in (7), we define

$$c(E_{31}) = \begin{pmatrix} 0 & & & & & \lambda_2 \\ \lambda_2 & 0 & & & & \\ & \lambda_2 & \ddots & & & \\ & & \ddots & 0 & & \\ 0 & & & \lambda_2 & 0 & \end{pmatrix}.$$

Hence $\mathrm{rank}(c(E_{31}) - E_{31}) = z(c(E_{31}) - E_{31}) = 1$. For the other E_{ij} in (8), we simply define $c(I) = I$ and $c(\mathrm{Diag}(0, \cdots, 0, 1)) = O$.

As another example, for automata with batch arrivals, the transition matrix E will be of the form

$$(17) \qquad E = \begin{pmatrix} 0 & \mu & & & & & 0 \\ \lambda_1 & 0 & 2\mu & & & & \\ \vdots & \lambda_1 & 0 & \ddots & & & \\ \lambda_b & \vdots & \ddots & \ddots & s\mu & & \\ & \lambda_b & \ddots & \ddots & 0 & s\mu & \\ & & \lambda_b & \ddots & \lambda_1 & 0 & s\mu \\ 0 & & & r_b & \cdots & r_1 & 0 \end{pmatrix},$$

where b is the largest possible batch size, λ_j is the arrival rate of batches with size j and $r_j = \sum_{i=j}^{b} \lambda_i$ for $j = 1, \ldots, b$. Here we define

$$c(E) = \begin{pmatrix} 0 & s\mu & & 0 & \lambda_b & \cdots & \lambda_1 \\ \lambda_1 & 0 & s\mu & & \ddots & \ddots & \vdots \\ \vdots & \lambda_1 & 0 & \ddots & & \ddots & \lambda_b \\ \lambda_b & \vdots & \ddots & \ddots & s\mu & & 0 \\ & \lambda_b & \ddots & \ddots & 0 & s\mu & \\ 0 & & \lambda_b & \ddots & \lambda_1 & 0 & s\mu \\ s\mu & 0 & & \lambda_b & \cdots & \lambda_1 & 0 \end{pmatrix}.$$

In this case $\mathrm{rank}(E - c(E)) = \max(s, b) + 1$ and $z(E - c(E)) = s + b$.

With these examples in mind, we are now ready to define our circulant approximations.

DEFINITION 1. *For $i = 1, \ldots, m$, $c(E_{i1})$ is defined to be the circulant matrix such that (i) the number of nonzero columns of $E_{i1} - c(E_{i1})$ is a constant less than l_1 and is independent of l_1, i.e.*

$$(18) \qquad z(E_{i1} - c(E_{i1})) = \wp(l_1),$$

and (ii) $c(E_{i1})$ is a nonnegative matrix, i.e.

$$(19) \qquad c(E_{i1}) \geq O.$$

We remark that requirements (18) and (19) are very general. The examples of $c(E_{i1})$ given above and also those in §6 all satisfy these two requirements provided that the queueing parameters such as s_i, λ_i, μ_i and b ($< l_1 - s$) are all independent of l_1. In general, there are many different forms of E_{i1} depending on the SAN itself. However, in many applications of the SANs [**4, 5, 6, 16, 15, 20, 21, 22, 31**], the transition matrix E_{11} of the main automaton takes the form of E_{ii} in (6), which is the transition matrix of an (M/M/s_i/l_i) queue. It has been shown in [**30, 33**] that the time spent (service time) in a jobshop is asymptotically exponentially distributed. Thus we may approximate a complex automaton with Poisson arrival process by E_{ii} of the form in (6) and $c(E_{ii})$ so defined in (16) will satisfy (18) and (19).

Next we define the circulant approximations $c(D_{i1})$ for the diagonal matrices D_{i1}. Similar to (10), we define the diagonal entries of $c(D_{i1})$ to be the column sums of $c(E_{i1})$, which are the same for all columns, as $c(E_{i1})$ are circulant matrices.

DEFINITION 2. *For $i = 1, \ldots, m$, $c(D_{i1})$ is defined to be the constant diagonal matrix such that*

$$(20) \qquad \mathbf{1}^t c(D_{i1}) = \mathbf{1}^t c(E_{i1}), \quad i = 1, \ldots, m.$$

Recall that the diagonal entries of D_{i1} are the column sums of E_{i1}. Hence by (18), we see that

$$(21) \qquad z(D_{i1} - c(D_{i1})) = \emptyset(l_1), \quad i = 1, \ldots m.$$

To construct our approximation, let us first define (cf (9))

$$(22) \qquad c(A) = \sum_{i=1}^{m} \left\{ c(D_{i1}) \bigotimes_{j=2}^{n} D_{ij} - c(E_{i1}) \bigotimes_{j=2}^{n} E_{ij} \right\},$$

We claim that $c(A)$ is irreducible.

PROPOSITION 3. *If the generator matrix $D_{ii} - E_{ii}$, $i = 1, \ldots, n$, of each stand-alone automaton is irreducible and that $c(E_{i1})$ satisfy (18) and (19) for $i = 1, \ldots, m$. Then $c(A)$ is irreducible. In particular, $c(A)$ has a one dimensional null space with positive null vectors.*

PROOF. By assumption, E_{ii} are irreducible for $i = 1, \ldots, n$. We claim that $c(E_{11})$ is irreducible. For if not, then since it is a circulant matrix, it can only be a constant diagonal matrix. Then by (18), E_{11} is the sum of a diagonal matrix and a matrix with $\emptyset(l_1)$ nonzero columns. Since $\emptyset(l_1) < l_1$, E_{11} cannot be irreducible, a contradiction. Thus $c(E_{11})$ is irreducible.

By the definition of $c(A)$ in (22) and the fact that the first n terms of A are given in (11), we see that the first n terms of $c(A)$ will be of the form

$$(c(D_{11}) - c(E_{11})) \otimes I \otimes \cdots \otimes I + I \otimes (D_{22} - E_{22}) \otimes I \otimes \cdots \otimes I + \cdots + I \otimes \cdots \otimes I \otimes (D_{nn} - E_{nn})$$

which is clearly irreducible. In particular, $\sum_{i=1}^{n} \{ c(E_{i1}) \bigotimes_{j=2}^{n} E_{ij} \}$ is also irreducible. Since E_{ij} and $c(E_{ij}) \geq O$ for all i and j, $\sum_{i=1}^{m} \{ c(E_{i1}) \bigotimes_{j=2}^{n} E_{ij} \}$ and hence $c(A)$ are irreducible too. By applying Proposition 2 to $c(A)$, we see that $c(A)$ has a one dimensional null space with positive null vectors. $\square$

Since $c(A)$ is singular, we cannot use it as a circulant approximation. Our approximation is constructed by perturbing $c(A)$ by a rank one matrix, similar to what we did in (14). In order to do it systematically, let us first look closely to the eigenvalues of $c(E_{i1})$ and $c(D_{i1})$. Recall that $c(D_{i1})$ are constant diagonal matrices, thus we may write

$$c(D_{i1}) = d_i I, \qquad i = 1, \ldots, m.$$

PROPOSITION 4. *Let F be the Fourier matrix of size l_1, i.e. the (j, k)th entry of F is given by $\exp(2\pi\sqrt{-1}jk/l_1)/\sqrt{n}$. For $i = 1, \ldots, m$, $c(E_{i1})$ can be diagonalized by F:*

(23) $$F^* c(E_{i1}) F = \mathrm{Diag}(t_{i1}, t_{i2}, \cdots, t_{il_1}).$$

Moreover,

(24) $$t_{il_1} = d_i, \quad i = 1, \ldots, m.$$

PROOF. Equation (23) follows from the fact that any circulant matrices can be diagonalized by the Fourier matrix of the same size, see [24]. To get (24), we first note that the last column of F is $\frac{1}{\sqrt{n}}\mathbf{1}$, i.e. $F\mathbf{e} = \frac{1}{\sqrt{n}}\mathbf{1}$, where $\mathbf{e} = (0, 0, \cdots, 0, 1)^t$. Thus by (20), we have

$$t_{il_1} = \mathbf{e}^t F^* c(E_{i1}) F = \frac{1}{\sqrt{n}}\mathbf{1}^t c(E_{i1}) F = \frac{1}{\sqrt{n}}\mathbf{1}^t c(D_{i1}) F = \frac{d_i}{\sqrt{n}}\mathbf{1}^t F = d_i, i = 1, \ldots, m.$$

$\square$

Using (23), we then have

(25) $$(F^* \otimes I) c(A)(F \otimes I)$$

$$= \sum_{i=1}^{m}\left\{ d_i I \bigotimes_{j=2}^{n} D_{ij} - \mathrm{Diag}(t_{i1}, t_{i2}, \cdots, t_{il_1}) \bigotimes_{j=2}^{n} E_{ij} \right]$$

$$= \mathrm{Diag}\left[\sum_{i=1}^{m}\left\{ d_i \bigotimes_{j=2}^{n} D_{ij} - t_{i1} \bigotimes_{j=2}^{n} E_{ij} \right\}, \cdots, \sum_{i=1}^{m}\left\{ d_i \bigotimes_{j=2}^{n} D_{ij} - t_{il_1} \bigotimes_{j=2}^{n} E_{ij} \right\} \right],$$

which is a diagonal block matrix. Using (24), the last diagonal block in (25) becomes

$$\sum_{i=1}^{m} d_i \left\{ \bigotimes_{j=2}^{n} D_{ij} - \bigotimes_{j=2}^{n} E_{ij} \right\}.$$

Premultiplying $\mathbf{1}^t$ to this matrix we get,

$$\mathbf{1}^t \cdot \sum_{i=1}^{m} d_i \left\{ \bigotimes_{j=2}^{n} D_{ij} - \bigotimes_{j=2}^{n} E_{ij} \right\} = \sum_{i=1}^{m} d_i \left\{ \mathbf{1}^t \cdot \bigotimes_{j=2}^{n} D_{ij} - \mathbf{1}^t \cdot \bigotimes_{j=2}^{n} E_{ij} \right\} = \mathbf{0}^t,$$

where the last equality follows from (10). Thus the last diagonal block in (25) is a singular matrix.

Since by Proposition 3, $c(A)$ has only a one-dimensional null space, the last diagonal block in (25) is the only singular block. All the other diagonal blocks in (25) are nonsingular. Similar to the proof in Proposition 1, we can easily prove that this last diagonal block is an irreducible matrix. Hence by Proposition 2, it also has a one dimensional null space with positive null vectors. To get our nonsingular approximation, we replace this last block by a nonsingular matrix using a rank one perturbation, as we did in (14):

$$H \equiv \left[\sum_{i=1}^{m} d_i \left\{ \bigotimes_{j=2}^{n} D_{ij} - \bigotimes_{j=2}^{n} E_{ij} \right\} + \mathbf{e}\mathbf{e}^t \right].$$

Thus our approximation is defined as

$$C \equiv (F \otimes I)\mathrm{Diag} \left[\sum_{i=1}^{m} \left\{ d_i \bigotimes_{j=2}^{n} D_{ij} - t_{i1} \bigotimes_{j=2}^{n} E_{ij} \right\}, \cdots, \right.$$

$$(26) \qquad \left. \sum_{i=1}^{m} \left\{ d_i \bigotimes_{j=2}^{n} D_{ij} - t_{i(l_1-1)} \bigotimes_{j=2}^{n} E_{ij} \right\}, H \right] (F^* \otimes I).$$

By the above arguments, C is clearly nonsingular. Moreover,

$$(27) \qquad \mathrm{rank}(c(A) - C) = 1.$$

In the following, we will use C in (26) to solve the linear system (14) through the Sherman-Morrison formula [**26**, p.3]. The cost depends on the cost of the matrix-vector multiplications of the form $G\mathbf{u}$ and also the cost of solving $C\mathbf{v} = \mathbf{u}$. In multiplying $G\mathbf{u}$, we can make use of the tensor structure of A as given in (9) and also the special structure of the transition matrices E_{ij}. Usually, E_{ij} are either sparse or near-Toeplitz matrices, cf. (5)–(8) and (17). The cost is therefore either of order $O(l_1)$ or $O(l_1 \log l_1)$.

The main cost for solving the approximation system $C\mathbf{v} = \mathbf{u}$ comes from (i) the matrix-vector multiplications by the Fast Fourier Transform (see (26)) and (ii) solving the diagonal block systems in (26). The cost for (i) is of $O(l_1 \log l_1)$. The cost for (ii) depends on the structure of the individual blocks in (26) which are of size $\prod_{i=2}^{n} l_i$-by-$\prod_{i=2}^{n} l_i$. Again fast algorithms for solving the block systems should make use of the sparse or near-Toeplitz structure of the blocks. In any case, the cost of solving each block system will be independent of l_1. As there are l_1 diagonal blocks in (26), the total cost will be of order $O(l_1)$. Hence the total cost for solving the approximation system is of $O(l_1 \log l_1 + l_1)$. Clearly, one can speed up (ii) by solving the diagonal block systems in parallel.

5. Practical Examples

In this section, we consider three practical SANs and discuss the computational cost for of our method discussed in §4. Recall that the main cost of our method comes from the matrix-vector multiplications of the form $C^{-1}\mathbf{y}$. The examples come from queueing systems, communication systems and manufacturing systems. In the following, we let N be the size of the generator matrix, C be the cost of the matrix-vector multiplications of the form $C^{-1}\mathbf{y}$ and T be the total cost of solving the steady state probability vector.

5.1. Overflow Queueing Systems.
We first consider the 2-queue overflow networks discussed in §2 where overflow is permitted from queue 2 to queue 1 when queue 2 is full. Thus the performance of queue 1 is important. We are interested in finding the steady state distribution vector when the queue length of the first queue increases.

We remark that if the Sherman-Morrison formula [**26**, p.3] is used, then we need $(s_1 + 1)l_2 + 1$ matrix-vector multiplications of the form $C^{-1}\mathbf{y}$. The following table shows the costs of M, C and T.

M	C	T
$l_1 l_2$	$O(l_1 \log l_1)$	$O(s_1 l_2 l_1 \log l_1)$

Table 1.

5.2. Telecommunication Systems.

In this section, we present an $(\mathrm{MMPP}/\mathrm{M}/s/s+m)$ network arising in telecommunication [**29**]. A Markov Modulated Poisson Process (MMPP) is a Poisson process whose instantaneous rate varies according to an irreducible Markov chain, see for instance [**29**]. Let us first define the system parameters, see Figure 2. We let $1/\lambda$ be the mean arrival time of the exogenously originating calls of the main queue, $1/\mu$ the mean service time of each server of the main queue, s the number of servers in the main queue, $l - s - 1$ the number of waiting spaces in the main queue, n the number of overflow queues, and finally (Q_j, Λ_j), $1 \le j \le n$, the parameters of the MMPP's modeling overflow parcels, where

$$Q_j = \begin{pmatrix} \sigma_{j1} & -\sigma_{j2} \\ -\sigma_{j1} & \sigma_{j2} \end{pmatrix} \quad \text{and} \quad \Lambda_j = \begin{pmatrix} \lambda_j & 0 \\ 0 & 0 \end{pmatrix}.$$

Here σ_{j1}, σ_{j2} and λ_j, $1 \le j \le n$, are positive MMPP parameters.

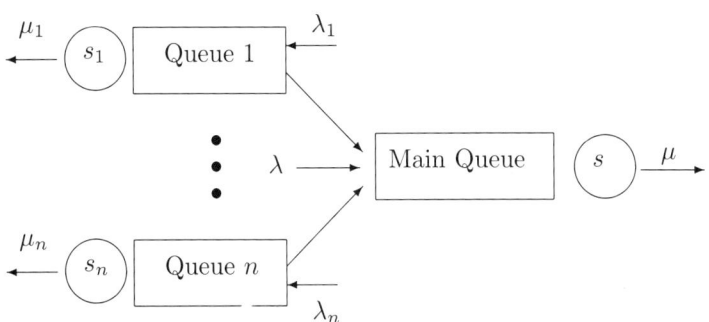

Figure 2. The Telecommunication System.

The input of the main queue comes from its own exogenous arrivals and the superposition of several independent MMPPs, which is still an MMPP and is parameterized by two $2^n \times 2^n$ matrices (Q, Γ). Here

$$Q = (Q_1 \otimes I_2 \otimes \cdots \otimes I_2) + (I_2 \otimes Q_2 \otimes I_2 \otimes \cdots \otimes I_2) + \cdots + (I_2 \otimes \cdots \otimes I_2 \otimes Q_n),$$

$$\Lambda = (\Lambda_1 \otimes I_2 \otimes \cdots \otimes I_2) + (I_2 \otimes \Lambda_2 \otimes I_2 \otimes \cdots \otimes I_2) + \cdots + (I_2 \otimes \cdots \otimes I_2 \otimes \Lambda_n)$$

and $\Gamma = \Lambda + \lambda I_{2^n}$, where I_2 and I_{2^n} are the 2×2 and $2^n \times 2^n$ identity matrices respectively.

We can regard the $(\mathrm{MMPP}/\mathrm{M}/s/l)$ queue as a Markov process on the state space

$$\{(i, j) \mid 0 \le i \le l - 1, 1 \le j \le 2^n\}.$$

The number i corresponds to the number of calls at the destination, while j corresponds to the state of the Markov process with generator matrix Q. Hence the generator matrix of the queueing process is given by the following $l2^n \times l2^n$ tridiagonal block matrix:

$$A = \begin{pmatrix} Q+\Gamma & -\mu I & & & & & 0 \\ -\Gamma & Q+\Gamma+\mu I & -2\mu I & & & & \\ & \ddots & \ddots & \ddots & & & \\ & & -\Gamma & Q+\Gamma+s\mu I & -s\mu I & & \\ & & & \ddots & \ddots & \ddots & \\ & & & & -\Gamma & Q+\Gamma+s\mu I & -s\mu I \\ 0 & & & & & -\Gamma & Q+s\mu I \end{pmatrix}.$$

It can be rewritten as

$$A = I_l \otimes Q + P_1 \otimes I_{2^n} + P_2 \otimes \Gamma.$$

where P_1 and P_2 take the form of Q_i and R in (2) and (3) respectively.

We note that in this example there are $(n+1)$ individual automata and $m = (2n+2)$ probabilistic rules. When Sherman-Morrison formula [**26**, p.3] is applied, then we need $(s+1)2^n$ matrix-vector multiplications of the form $C^{-1}\mathbf{y}$ and the cost for the matrix-vector multiplication $C^{-1}\mathbf{y}$ is of $O(2^n l \log l)$, see [**15**]. The following table shows the costs of M, C and T.

M	C	T
$2^n l$	$O(2^n l \log l)$	$O(s2^{2n} l \log l)$

Table 2.

5.3. The Manufacturing System. In this subsection, we consider a manufacturing system of two machines in tandem under the hedging point product policy, see [**22**] and Figure 3. The system parameters are: $1/\lambda$, the mean inter-arrival time of a demand; $1/\mu_1$, the mean unit processing time of the first machine; $1/\mu_2$, the mean unit processing time of the second machine; b_1, the size of the buffer B_1 for the first machine; b_2, the maximum size of the buffer B_2 for the finished products; h, the hedging point; and m, the maximum allowable backlog. We note that the inventory level of the first buffer cannot be negative or exceed the buffer size b_1. Thus the total number of possible inventory levels in the first buffer is $(b_1 + 1)$. For the second buffer, under the hedging point policy, the maximum possible inventory level is h with $h \leq b_2$. Since we allow a maximum backlog of m in the system, the total number of possible inventory levels in the second buffer is $l = (m+h+1)$. In practice the value of l can easily go up to thousands.

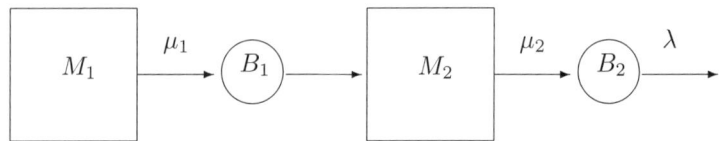

Figure 3. Two Machines in Tandem.

We let $z_1(t)$ and $z_2(t)$ be the inventory levels of the first and second buffers at time t respectively. Then $z_1(t)$ and $z_2(t)$ take integer values in $[0, b_1]$ and $[-m, h]$ respectively. Thus the joint inventory process $\{(z_1(t), z_2(t)), t \geq 0\}$ is a continuous time Markov chain taking values in the state space

$$S = \{(z_1(t), z_2(t)) : z_1 = 0, \cdots, b_1,\ z_2 = -m, \cdots, h.\}.$$

We order the inventory states lexicographically, with $z_1(t)$ first and then $z_2(t)$. Then we obtain the tridiagonal block generator for the joint inventory system

$$A = \begin{pmatrix} \Lambda + \mu_1 I_l & \Sigma & & & 0 \\ -\mu_1 I_l & \Lambda + D + \mu_1 I_l & \Sigma & & \\ & \ddots & \ddots & \ddots & \\ & & -\mu_1 I_l & \Lambda + D + \mu_1 I_l & \Sigma \\ 0 & & & -\mu_1 I_l & \Lambda + D \end{pmatrix},$$

where

$$\Lambda = \begin{pmatrix} 0 & -\lambda & & 0 \\ \lambda & \ddots & \ddots & \\ & \ddots & & -\lambda \\ 0 & & \lambda \end{pmatrix}, \quad \Sigma = \begin{pmatrix} 0 & & & 0 \\ -\mu_2 & \ddots & & \\ & \ddots & \ddots & \\ 0 & & -\mu_2 & 0 \end{pmatrix},$$

and $D = \mathrm{Diag}(\mu_2, \cdots, \mu_2, 0)$ is an $l \times l$ diagonal matrix. We note that A can be written as

(28) $\qquad A = I_{b_1+1} \otimes \Lambda + W_1 \otimes I_l + \mathrm{Diag}(0, 1, \cdots, 1) \otimes D + W_2 \otimes \Sigma,$

where

$$W_1 = \begin{pmatrix} \mu_1 & 0 & & 0 \\ -\mu_1 & \mu_1 & & \\ & \ddots & \ddots & \\ 0 & & -\mu_1 & 0 \end{pmatrix} \quad \text{and} \quad W_2 = \begin{pmatrix} 0 & 1 & & 0 \\ & \ddots & \ddots & \\ & & \ddots & 1 \\ 0 & & & 0 \end{pmatrix}.$$

Similar to the discussion in §4, our approximation is obtained by taking the circulant approximations of the matrices Λ, Σ and D in (28). It is easy to check that our approximation is unitary similar to a diagonal block matrix with each block being a tridiagonal matrix. The cost for the matrix-vector multiplication $C^{-1}\mathbf{y}$ is of $O(l \log l)$. For this problem, when Sherman-Morrison formula is used, we need $2(b_1 + 1) + 1$ matrix-vector multiplications of the form $C^{-1}\mathbf{y}$. The following table shows the costs of M, C and T when l.

N	C	T
lb_1	$O(l \log l)$	$O(b_1 l \log l)$

Table 3.

6. Concluding Remarks

In this paper, we discuss circulant approximations for stochastic automata new-torks. Our approximations are constructed by taking circulant approximations of the generator matrices of the networks. We discuss the application of our method for systems from three different applications.

We remark that our approximations can be applied to manufacturing systems of more than two machines (jobshops) in tandem, see [22] for instance. It will be interesting to extend our results to other sophisticated Markovian models [18, 23].

Acknowledgement

The authors would like to thank the referees for their helpful suggestions.

References

[1] M. Ajmone-Marsan, G. Balbo, G. Conte, S. Donatelli, and G. Franceschinis, *Modelling with Generalized Stochastic Petri Nets*, Wiley, New York, 1995.

[2] F. Baccelli et. al, *Synchronization and Linearity: An Algebra for Discrete-event Systems*, Wiley, New York, 1992.

[3] A. Berman and R. Plemmons, *Nonnegative Matrices in Mathematical Sciences*, SIAM, Philadelphia, 1994.

[4] P. Buchholz, *A Class of Hierarchical Queueing Networks and Their Analysis*, Queueing Systems, 15 (1994), pp. 59–80.

[5] P. Buchholz, *Hierarchical Markovian Models: Symmetries and Aggregation*, Performance Evaluation, 22 (1995), pp. 93–110.

[6] P. Buchholz, *Equivalence Relations for Stochastic Automata Networks*, Computations of Markov Chains: Proc. of the 2nd International Workshop On Numerical Solutions of Markov Chains, Kluwer, 1995, pp. 197–216.

[7] J. Buzacott and J. Shanthikumar, *Stochastic Models of Manufacturing Systems*, Prentice-Hall International Editions, New Jersey, 1993.

[8] R. Chan, *Iterative Methods for Overflow Queueing Models I*, Numer. Math., 51 (1987), pp. 143–180.

[9] R. Chan, *Iterative Methods for Overflow Queueing Models II*, Numer. Math., 54 (1988), pp. 57–78.

[10] R. Chan, *Iterative Methods for Queueing Networks with Irregular State-Spaces*, Proceedings to the IMA Workshop on Linear Algebra, Markov Chains and Queueing Models, Minneapolis, January 13–17, 1992, pp. 89–110, Eds: C. Meyer and R. Plemmons, Springer-Verlag, 1993.

[11] R. Chan and W. Ching, *Toeplitz-circulant Preconditioners for Toeplitz Systems and Their Applications to Queueing Networks with Batch Arrivals*, SIAM J. Sci. Comput., 17 (1996), pp. 762–772.

[12] R. Chan and W. Ching, *Circulant Preconditioners for Stochastic Automata Networks*, to appear in Numer. Math.

[13] R. Chan and K. Ng, *Conjugate Gradient Methods for Toeplitz Systems*. SIAM Review, 38 (1996), pp. 427–482.

[14] R. Chan, C. Wong and W. Ching *Optimal Trigonometric Preconditioners for Elliptic Problems and Queueing Problems*, SEA Bull. Math., 3 (1996), pp. 117–124.

[15] W. Ching, R. Chan and X. Zhou, *Circulant Preconditioners for Markov Modulated Poisson Processes and Their Applications to Manufacturing Systems*, SIAM J. Matrix Anal. Appl., 17 (1997), pp. 452–467.

[16] W. Ching and X. Zhou, *Matrix Methods for Production Planning in Failure Prone Manufacturing Systems*, Lecture Notes in Control and Information Sciences, Springer-Verlag, 1996, pp. 3–30.

[17] W. Ching and X. Zhou, *Optimal (S,s) Production Policies with Delivery Time Guarantee*, Lectures in Applied Mathematics, Mathematics of Stochastic Manufacturing Systems, The American Mathematical Society, 1997, pp. 71–82.

[18] W. Ching and X. Zhou, *Machine Repairing Models for Manufacturing Systems*, The 5th IEEE Conference on Emerging Technologies and Factory Automation, November, 1996.

[19] W. Ching and X. Zhou, *Circulant Approximation for Preconditioning in Stochastic Automata Networks*, to appear in Comp. Math. Appl.

[20] W. Ching, *Circulant Preconditioners for Failure Prone Manufacturing Systems*, Lin. Alg. Appl., 266 (1997), pp. 161–180.

[21] W. Ching, *Markov Modulated Poisson Processes for Multi-location Inventory Problems*, Inter. J. Prod. Econ., 53 (1997), pp. 232–239.

[22] W. Ching, *Iterative Methods for Manufacturing Systems of Two Stations in Tandem*, Inter. J. Appl. Math. Letters, 11 (1998), pp. 7–12.

[23] W. Ching, *Optimal (S,s) Policies for Manufacturing Systems of Unreliable Machines in Tandem*, International Symposium on Product Quality and Integrity, Anaheim, U.S.A., January, 1998.

[24] P. Davis, *Circulant Matrices*, John Wiley and Sons, New York, 1979.

[25] P. Glasserman and D. Yao, *Monotone Structure in Discrete-event Systems*, Wiley, New York, 1994.

[26] G. Golub and C. Van Loan, *Matrix Computations*, 2nd ed., John Hopkins University Press, Baltimore, 1984.

[27] H. Heffes and D. Lucantoni, *A Markov Modulated Characterization of Packetized Voice and Data Traffic and Related Statistical Multiplexer Performance*, IEEE J. Select. Areas Commun., SAC-4 (1986), pp. 856–868.

[28] L. Kaufman, *Matrix Methods for Queueing Problems*, SIAM J. Sci. Statist. Comput., 4 (1982), pp. 525–552.

[29] K. Meier-Hellstern, *The Analysis of a Queue Arising in Overflow Models*, IEEE Trans. Commun., 37 (1989), pp. 367–372.

[30] H. Mendelson and S. Whang, *Optimal Incentive-Compatible Priority Pricing for the M/M/1 Queue*, Operat. Res., 38 (1990), pp. 870–883.

[31] B. Plateau and K. Atif, *Stochastic Automata Network for Modelling Parallel Systems*, IEEE Tran. Software Engineering, 12 (1991), pp. 370–389.

[32] Y. Saad, *Iterative Methods for Sparse Linear Systems*, PWS Publishing Co., Boston, 1996.

[33] P. Shanthikumar, S. Datar and R. Akella, *Approximations for the time spent in a dynamic job shop with applications to due date assignment*, Inter. J. Prod. Res., 26 (1988), pp. 1329–1352.

[34] W. Stewart, K. Atif and B. Plateau, *The Numerical Solution of Stochastic Automata Networks*, Euro. J. Operat. Res., 86 (1995), 503–525.

[35] H. Young, B. Byung and K. Chong, *Performance Analysis of Leaky-bucket Bandwidth Enforcement Strategy for Bursty Traffics in an ATM Network*, Comput. Net. ISDN Syst., 25 (1992), pp. 295–303.

AMS/IP Studies in Advanced Mathematics
Volume 26, 2002

Estimating the Speed of Random Walks

Dayue Chen

ABSTRACT. In this survey paper we discuss the recent developments in estimating the speed of random walks on trees and general graphs.

1. Introduction

Lattice Z^d is the default choice of state space for discrete stochastic processes. Random walks and percolation theory on Z^d are classical examples. Alternatively one may replace Z^d by a genereal graph. Studies of probability models on general graphs started in the 1950's, see Kesten (1959) and Grenander (1963), and boomed in late 80's. For example, the paper by Pemantle (1992) set off an intensive investigation on the contact process on trees. Probability on trees and networks is now studied by more people. A comprehensive book is being written by R. Lyons and Y. Peres (1999). Also see the survey paper by Peres (1997).

For graph G, we denote by $V(G)$ and $E(G)$ respectively the sets of vertices and edges of G. The graphs on which probability models are defined are mainly the following three classes: 1) trees, 2) transitive or semi-transitive graphs, 3) random graphs. Trees are connected graphs without cycles. There is a unique path connecting any pair of vertices. The Cayley graph of a finitely generated group is a typical example of transitive graphs. Suppose that group G is generated by a finite set $A = \{a_1, a_2, \cdots, a_n\}$ and that A is symmetric, *i.e.* that $a_i \in A$ implies that $a_i^{-1} \in A$. Then vertices of the corresponding graph are the elements of the group. There is an edge connecting vertices u and v if and only if $u^{-1}v \in A$. Lattice Z^d and homogeneous tree T_d are two examples of Cayley graphs. Galton-Watson trees and random clusters in a bond percolation are nice examples of random graphs. We assume that readers are familiar with these random graphs. Otherwise readers may consult references such as Athreya and Ney (1972) and Grimmett (1989). When we are talking about random graphs, we are actually talking about a probability measure on a family of graphs. We are interested in properties that are valid for all samples almost surely with respect to that probability measure.

The simplest probability model is random walks. It is a corner stone to study other more comlex models. A *simple random walk* is defined as a Markov chain on $V(G)$ with transition probability $p(u, v) = 1/d_u$ if $v \sim u$, and 0 otherwise. Here

$v \sim u$ means that there is an edge connecting u and v, and d_u is the degree of u. With an initial state (or distribution), a random walk is well defined.

There are, of course, many other versions of random walks. For tree T we may define λ-*biased random walk* as follows. A vertex is selected as the *root* and is denoted by o. The distance $|v|$ from vertex v to o is also called the *level* or *generation* of v. For vertex v of level $|v|$, there is a unique adjacent vertex which is of level $|v| - 1$. This unique adjacent vertex is called the *parent* of v, and is denoted by v_*. Other adjacent vertices of v are all of level $|v| + 1$, and are called *children* of v. Let k_v be the number of children of v. It is also known as the *branching number* of v. Children of v are denoted by v_i, $i = 1, 2, \cdots k_v$. For positive number λ, define λ-biased random walk $\{X_n\}$ as a Markov chain on $V(T)$ with transition probability

$$(1) \qquad p(v, v_*) = \frac{\lambda}{\lambda + k_v}, \qquad p(v, v_i) = \frac{1}{\lambda + k_v}, \quad v \neq o.$$

The transition probability at o is different slightly in accordance with the lack of o_*. Let k_o be the branching number of o and o_i is a child of o. Define $p(o, o_i) = 1/k_o$.

Random walks are classified as recurrent or transient, as in the case of Z^d. It turns out that this probabilistic property is intimately related to geometric properties of the graph. See Kesten (1959), Kaimanovich and Vershik (1983). Similarly, transience/recurrence of the Brownian motion on a manifold is related to geometric properties of the manifold, see Grigor'yan(1998). For the discrete case there are now some efficient criteria, for example, Griffeath and Liggett(1982) and Lyons(1983), to determine if a random walk is recurrent or transient. Applying the criteria, it is not difficult to obtain the dichotomy of recurrence and transience.

Theorem 1 (Lyons (1990), Lyons (1992)). *The λ-biased random walk on tree T is transient if and only if the branching number $br(T) > \lambda$.*

The definition of *branching number* $br(T)$ of tree T is somewhat involved and is deferred to the next section.

Transient random walks are further studied by examining the *speed*. Let X_n be the position of the walker after n movements from the initial vertex o, and $|X_n|$ be the graphic distance (the least number of edges) from o to X_n. Whenever it exists, $\lim_{n \to \infty} |X_n|/n$ is called the *speed,* or the *rate of escape to infinity,* of the random walk $\{X_n\}$ starting at o. Similarly, $\liminf_{n \to \infty} |X_n|/n$ is called the *lower speed* of random walk $\{X_n\}$.

2. Random Walks on Trees

Trees are the simplest graphs. An m-ary tree is a rooted tree that each vertex has m "children". In particular it is called a binary tree if $m = 2$. Obviously the speed of λ-biased random walk on an m-ary tree exists and is $(m - \lambda)/(m + \lambda)$ if $m \geq \lambda$. Next, let us consider the *Fibonacci tree.* There are two types of vertices. A type 1 vertex has one child of type 2, while a type 2 vertex has two children, one of type 1 and one of type 2. It is calculated in Lyons, Pemantle and Peres (1997) that

$$Speed(\lambda) = \frac{(\sqrt{\lambda + 1} + 2)(\sqrt{\lambda + 1} - \lambda)}{\sqrt{\lambda + 1}(2 + \lambda + \sqrt{\lambda + 1})}$$

for $\lambda \leq (\sqrt{5}+1)/2$. The recursive argument can be also used for periodic trees with N types of vertices. See Takacs (1997).

In general, however, the speed can not be always identified expressively. To estimate the speed, we first introduce the definition of the *branching number* of a tree. Choose a vertex o as the root. Define the distance $|v|$ from vertex v to o to be the (minimal) number of edges of a path connecting v to o. Similarly $|e|$ is defined for edge e. A subset $\Pi \subset E$ is called a *cutset* if every path from o to infinity will intersect Π. Define the branching number

$$br(T) = \inf\{\alpha, \quad \inf_{\Pi} \sum_{e \in \Pi} \alpha^{-|e|} < \infty\}.$$

Certainly, the definition is independent of the choice of o. For a homogeneous tree, namely, every vertex has the same number of edges, say $d+1$, then $br(T) = d$. The branching number of the Fibonacci tree is $(1 + \sqrt{5})/2$. Tree T is called a *Galton-Watson tree* if it is a realization of a *Galton-Watson process*. A *Galton-Watson process* is determined by a probability distribution $\{p_0, p_1, p_3, \cdots\}$ on non-negative integers. Note that the degree of root o is k_o, and the degree of v other than root is $1 + k_v$, k_v's are *i.i.d.* random variables with the probability distribution $P(k_v = j) = p_j$. Let $m = \sum_j jP(k_v = j)$ be the mean number of offsprings. Then for Galton-Watson trees $br(T) = m$ *a.s.*

For comparison we like to mention the *growth number*

$$gr(T) = \liminf_n |B(o,n)|^{1/n}.$$

Here $B(o,n) = \{v \in V(T); |v| \leq n\}$ is the ball of vertices centered at o with radius n. $|B|$ is the cardinality of set B. While both numbers characterize how fast a tree grows, the growth number only counts the number of vertices and ignores the graphic structure of tree. The branching number does a better job. Indeed we have examples that $gr(T) > 1$ and $br(T) = 1$.

Theorem 2 (Virag (1998)). *For λ-biased random walk on tree T*

$$\liminf_{n \to \infty} \frac{|X_n|}{n} \leq \frac{br(T) - \lambda}{br(T) + \lambda}.$$

On the other hand, Y. Peres provided a lower bound on the branching number $br(T)$ if the speed of simple random walk is positive.

Theorem 3 *If $\liminf_n |X_n|/n \geq s > 0$, then*

$$br(T) \geq \frac{(1+s)\log(1+s) + (1-s)\log(1-s)}{2}.$$

Next, we discuss the λ-*biased random walk* on Galton-Watson trees which is defined as follows. Assume that the probability law of offspring k_v satisfies

$$\sum_{i=1}^{\infty} P(k = i) = 1, \qquad m = \sum_i iP(k = i) < \infty.$$

The assumption $P(k = 0) = 0$ implies that the Galton-Watson process is supercritical. Let $\mathcal{T}$ be the collection of all rooted trees. The probability law of k_v induces naturally a probability measure in $\mathcal{T}$. First, take a Galton-Watson tree T

according to the induced probability measure in $\mathcal{T}$. Then, define a random walk $\{X_n\}$ on T with transition probability given in (1). Here λ is a positive number. Thus a point in the big probability space has two components: a random tree and a random path. The probability law of k_v's and parameter λ determine jointly a unique probability measure in this big space. It is easy to conlude by the same argument of Theorem 1 that the random walk on Galton-Watson trees is transient *a.s.* in the *big* space if $m > \lambda$. A more profound finding is the existence of the *speed* of the random walk.

Theorem 4 (Lyons, Pemantle and Peres (1996a)). *For a fixed λ ($\lambda < m$) and for a.e. all Galton-Watson trees, $\lim_{n\to\infty} |X_n|/n$ exists and the limit is a positive constant depending only on λ and the offspring distribution.*

The key point of the proof is to show that regenerating points are independently and identically distributed. A vertex is called a *regenerating point* of a path of random walk if it is visited by the walker exactly once. The Strong Law of Large Numbers is then applied to the distances and the times between consective regenerating points to conclude the claim of Theorem 4. For the case $\lambda = 1$, they computed the speed explicitly in Lyons, Pemantle and Peres (1995).

$$\lim_{n\to\infty} \frac{|X_n|}{n} = \sum_{i=1}^{\infty} P(k = i)\frac{i-1}{i+1}.$$

Assumption $P(k = 0) = 0$ can be removed in this case but the expression will be much more complicated. These results lead Lyons, Pemantle and Peres to believe that

$$\lim_{n\to\infty} \frac{|X_n|}{n} \leq \frac{m - \lambda}{m + \lambda} \qquad a.s.$$

The last inequality was first formulated in Lyons, Pemantle and Peres (1997) as a conjecture and inspired many sebsequent research works in this subject. A partial verification was first obtained by the author.

Theorem 5 (Chen (1996)). *If $P(k = 0) = 0$, $\lambda \leq 1$, then*

(2)
$$\frac{m' - \lambda}{m' + \lambda} \leq \lim_n \frac{|X_n|}{n} \leq \frac{m - \lambda}{m + \lambda}$$

where m' is the harmonic mean of branching numbers of Galton-Watson trees. Namely $1/m' = \sum_{j=1}^{\infty} P(k_v = j)/j$.

Note there is also a lower bound in (2) which is usually more difficult to establish in most cases. It would be interesting to provide a lower bound for all $\lambda \leq br(T)$. Also the following two problems still remain open.

Problem 6 (Lyons, Pemantle and Peres (1997)) *Is Speed(λ) monotone in λ? Is Speed(λ) a smooth function of λ?*

Some functional limit theorems for the range and for the speed of a simple random walk were obtained by D. Piau (1998).

3. Random Walks on Graphs

For $S \subset V(G)$ there is the maximal subgraph of G with S as its vertices. We say S is *connected* if the maximal subgraph is *connected*. Let

$$\iota_E(G) := \inf \left\{ \frac{|\partial_E S|}{|S|} : S \subset V(G),\ S \text{ is connected},\ 1 \leq |S| < \infty \right\}.$$

It is called the *Cheeger constant* or the *isoperimetric constant*. Here $|S|$ is the cardinality of S and $|\partial_E S|$ the number of boundary edges that one end is in S and the other one is not in S.

We say G is *amenable* if its Cheeger constant is zero, and *non-amenable* otherwise. We say G grows (sub-) exponentially if the volume of the ball of $V(G)$ centered at o and radius n grows (sub-) exponentially as $n \to \infty$. A graph that grows sub-exponentially is amenable. And a non-amenable graph grows exponentially.

G_1 and G_2 are amenable Cayley graphs with exponential growth. A vertex of G_1 can be identified as $(m, \eta) \in Z^1 \times \{\text{finite subsets of } Z^1\}$ and has three edges connecting it to one of the following neighboring vertices: $(m+1, \eta)$, $(m-1, \eta)$ and $(m, \eta \Delta \{m\})$. Here $\eta \Delta \{m\}$ is $\eta \setminus \{m\}$ if $m \in \eta$, and is $\eta \cup \{m\}$ if $m \notin \eta$. If one dimensional lattice Z^1 is replaced by two dimensional lattice Z^2 we get G_2.

Simple random walks on G_1 and G_2 were studied in Kaimanovich and Vershik (1983). The λ-biased random walk in G_1 was studied in Lyons, Pemantle and Peres (1996b). In that paper G_1 is called the *lamplighter group*. With the lamplighter interpretation, Z^1 or Z^2 is the sites of lamps, η is the set of lamps which are *on*, m is the position of the lamplighter. Each time the lamplighter either switches the lamp (from *on* to *off*, or from *off* to *on*) or moves to one of the nearest sites. It is interesting to note that the speed of λ-biased random walks on G_1 is not monotone in λ for $1 \leq \lambda \leq (1 + \sqrt{5})/2$. This observation makes Problem 6 suspensive.

A simple random walk in graph G has zero speed if G grows sub-exponentially. On the other hand, we have the following assertion concerning the positivity of speed.

Theorem 7. (Kesten (1959), Dodziuk (1984)) *The following three statements are equivalent for the simple random walk on a Cayley graph:*
   (1) the Cheeger constant is positive;
   (2) $\liminf_{n \to \infty} |X_n|/n$ *is positive;*
   (3) $\lim_{n \to \infty} (P(X_n = o))^{1/n} < 1$.

In *p-Bernoulli bond percolation* of G, each edge of G is independently declared *open* with probability p and *closed* with probability $1 - p$. The random subgraph consisting of $V(G)$ and all open edges is called a percolation of G. A connected component of the subgraph is called an *open cluster*. The fundamental problem of the percolation theory is to identify if there is an infinite open cluster. In a Cayley graph, there is a threshold p_c such that there is an infinite open cluster for $p > p_c$ and there is no infinite open cluster for $p < p_c$. Furthermore, Burton and Keane (1989) show that p-Burnoulli percolation has a.s. at most one infinite cluster when G is an amenable Cayley graph.

A simple random walk on an infinite open cluster of a percolation was first considered by G. Grimmett, H. Kesten and Y. Zhang in 1993. It is proved in their joint paper that a simple random walk on the infinite open cluster of a percolation of

Z^d is transient if $d \geq 3$. A simplified proof is now available in Benjamini, Pemantle and Peres (1997).

In some sense the property of transience is inherited in a percolation. One would like to ask if other properties are also inheritable. In particular, one likes to know if the positivity of speed of a simple random walk is inheritable. Certainly, any open cluster of G grows sub-exponentially if G does. On the other hand, positivity of the Cheeger constant and the translation invariance of a Cayley graph is totally lost in the percolation. An alternative is proposed in Benjamini, Lyons and Schramm (1998). Fix some basepoint $o \in V(G)$. Call

$$\iota_E^*(G) := \liminf_{n \to \infty} \left\{ \frac{|\partial_E S|}{|S|} : o \in S \subset V(G), S \text{ is connected}, n \leq |S| < \infty \right\}$$

the *anchored expansion constant* of G. Of course $\iota_E^*(G)$ is independent of the choice of basepoint o. It is hoped that $\iota_E^*(G)$ may replace the isoperimetric constant ι in the study of probability problems in G.

Conjecture 8 (Benjamini, Lyons and Schramm(1998), Conjecture 6.2) *Let G be a bounded degree graph with $\iota_E^*(G) > 0$. Then $\liminf_{n \to \infty} |X_n|/n$ is positive with positive probability.*

Since a subgraph of G can not grow faster than G does, a simple random walk in an infinite cluster of G has zero speed if G grows sub-exponentially. It is shown in Theorem 1.3 of Benjamini, Lyons and Schramm(1998) that a simple random walk in an infinite cluster has a positive speed if G is a non-amenable Cayley graph. The following two conjectures are proposed.

Conjecture 9 (Benjamini, Lyons and Schramm(1998), Conjectures 1.4 and 1.5). *If G is a Cayley graph on which simple random walk has positive speed, then a.s., simple random walk on each infinite cluster of p-Bernoulli percolation has positive speed.*

If G is a Cayley graph on which simple random walk has zero speed, then a.s., simple random walk on every cluster of Bernoulli percolation also has zero speed.

It is therefore worthwhile to study, as pointed out in Benjamini, Lyons and Schramm(1998), a simple random walk in the infinite open cluster of an amenable Cayley graph with exponential growth. As we explained before, G_1 and G_2 are amenable Cayley graphs with exponential growth.

Theorem 10 (Chen and Peres (1998)) *The simple random walk in an open cluster of the Bernoulli bond percolation of G_1 or G_2 has zero speed. Namely, $\lim_n |X_n|/n = 0$ a.s.*

The conclusion of Theorem 10 conforms with Conjecture 9. But it is just a special case. A generalization of this theorem is given in Theorem 3.1 of Chen and Peres(1998).

Acknowledgement. This is an expanded notes of the lecture delivered in the IMS Workshop on Applied Probability, May 31-June 11, 1999, at the Chinese University of Hong Kong. The author would like to thank the organizers for their hospitality. The preparation of this paper was upported in part by grants from the NSF of China, the Ministry of Education and the 863 Project.

References

[1] Athreya, K.B. and Ney P.E.(1972). *Branching Processes*. Springer-Verlag, New York.

[2] Benjamini, I., Lyons, R. and Schramm, O. (1998). *Percolation perturbation in potential theory and random walks*. Preprint.

[3] Benjamini, I.,Pemantle, R. and Peres, Y. (1997). *Unpredictable paths and percolation*. Preprint.

[4] Burton, R.M. and Keane, M. (1989). *Density and uniqueness in percolation,* Commun. Math. Phys. **121**, 501-505.

[5] Chen, D. (1997). *Average properties of random walks on Galton-Watson trees*. Ann. Inst. Henri Poincare – Probab. Statist. **33**, 359-369.

[6] Chen, D. and Peres, Y. (1998). *Anchored expansion, percolation and speed*. Preprint.

[7] Dodziuk, J. (1984) *Difference equations, isoperimetric inequality, and transience of certain random walks,* Trans. Amer. Math. Soc. **284**, 787-794

[8] Grigor'yan (1999). *Analytic and geometric background of recurrence and non-explosion of the Brownian motion on Riemannian manifolds*. Bulletin of Amer. Math. Soc. **36**, 135-249.

[9] Griffeath, D. and Liggett, T.M. (1982). *Critical phenomena for Spitzer's reversible nearest-particle systems,* Ann. Probab. **10**, 881-895.

[10] Grenander,U. (1963). *Probabilities on Algebraic Structures*. Wiley, New York.

[11] Grimmett, G. R. (1989). *Percolation,* Springer, New York.

[12] Grimmett, G., Kesten, H. and Zhang, Y. (1993). *Random walk on the infinite cluster of the percolation model*. Probab. Theory and Related Fields **96** 33-44.

[13] Kaimanovich, V. A. and Vershik, A. M. (1983). *Random walks on discrete groups: boundary and entropy,* Ann. Probab. **11**, 457–490.

[14] Kesten, H. (1959). *Symmetric random walks on groups*. Trans. AMS. **92** 336-354.

[15] Lyons, T. (1983). *A simple criterion for transience of a reversible Markov chain*. Ann. Probab. **11**, 393-402.

[16] Lyons, R. (1990). *Random walks and percolation on trees*. Ann. Probab. **18**, 931-958.

[17] Lyons, R. (1992). *Random walks, capacity and percolation on trees*. Ann. Probab. **20**, 2043-2088.

[18] Lyons, R., Pemantle, R. and Peres, Y. (1995). *Ergodic theory on Galton-Watson trees: speed of random walk and dimension of harmonic measure*. Ergod. Th. & Dynam. Sys., **15**, 1-27

[19] Lyons, R., Pemantle, R. and Peres, Y. (1996a). *Biased random walks on Galton-Watson trees*. Probab. Th. Rel. Fields, **106**, 249-264.

[20] Lyons, R., Pemantle, R. and Peres, Y. (1996b). *Random walks on the lamplighter group*. Ann. Probab. **24**, 1993–2006.

[21] Lyons, R., Pemantle, R. and Peres, Y. (1997). *Unsolved problems concerning random walks on trees*. In *Classical and Modern Branching Processes,* K. Athreya and P. Jagers (editors), pp223-238. Springer, New York.

[22] Lyons, R. and Peres, Y. (1999). Probability on Trees and Networks. In preparation. Current version available at http://www.ma.huji.ac.il/~lyons

[23] Peres, Y. (1997). *Probability on trees: an introductory climb*. Notes from St. Flour Summer School.

[24] Piau, D. (1998) *Théoreme central limite fonctionnel pour une marche au hasard en environnement aléatoire*. Ann. Probab., **26**, 1016-1040.

[25] Takacs, C. (1997). *Random walk on periodic trees*. Electronic J. of Prob. **2**, 1-16, available at http://www.math.washington.edu/~ejpecp/EjpVol2/

[26] Virag, B. (1998). *On the speed of random walks in trees*. Preprint.

[27] Virag, B. (1998). *On the anchored expansion property*. Preprint.

SCHOOL OF MATHEMATICAL SCIENCES, PEKING UNIVERSITY, BEIJING 100871, CHINA, , E-MAIL ADDRESS: DAYUE@PKU.EDU.CN

AMS/IP Studies in Advanced Mathematics
Volume 26, 2002

A new story of ergodic theory

Mu-Fa Chen

ABSTRACT. In the recent years, a great effort has been made to develop a new ergodic theory for Markov processes. It is mainly concerned with the study on several different inequalities. Some of them are very classical but some of them are rather new. The Liggett-Stroock form of Nash-type inequalities, the related ones and their comparison are discussed. Based on some new isoperimetric or Cheeger's constants, a simple sufficient condition for the inequalities is reported. The resulting condition can be sharp qualitatively. Finally, a diagram of the inequalities and the traditional three types of ergodicity is presented.

The paper is divided into twelve short sections. In the first eleven sections, various currently interested inequalities are discussed. The relationship of the inequalities and the three types of traditional ergodicity is exhibited in the last section, which may be glanced over before reading the details of the paper.

1. Notations

Let $(E, \mathcal{E}, \pi)$ be a measure space with σ-finite non-negative measure π. Denote by $L^p(\pi)$ the usual real L^p-space with norm $\|\cdot\|_p$. Consider a symmetric form D on $L^2(\pi)$ with domain $\mathcal{D}(D)$. Two typical forms are the following:

$$D(f) := D(f, f) = \tfrac{1}{2} \int_{\mathbb{R}^d} \langle a(x) \nabla f(x), \nabla f(x) \rangle \pi(\mathrm{d}x) + \int_{\mathbb{R}^d} c(x) f(x)^2 \pi(\mathrm{d}x),$$
$$\mathcal{D}(D) \supset C_0^\infty(\mathbb{R}^d),$$

where $\langle \cdot, \cdot \rangle$ is the standard inner product in $\mathbb{R}^d$, a is positive definite and $c \geq 0$;

$$(1) \qquad D(f) = \tfrac{1}{2} \int_{E \times E} J(\mathrm{d}x, \mathrm{d}y)[f(y) - f(x)]^2 + \int_E K(\mathrm{d}x) f(x)^2,$$
$$\mathcal{D}(D) = \{ f \in L^2(\pi) : D(f) < \infty \},$$

where J is a non-negative symmetric measure having no charge on the diagonal $\{(x, x) : x \in E\}$ and $K(\mathrm{d}x)$ is a non-negative measure. As usual, $D(f, g) := [D(f + g) - D(f - g)]/4$. A particular example in our mind is the symmetrizable jump process with q-pair $(q(x), q(x, \mathrm{d}y))$ and symmetrizing measure π, for which we have $J(\mathrm{d}x, \mathrm{d}y) = \pi(\mathrm{d}x) q(x, \mathrm{d}y)$. More especially, for a Q-matrix $Q = (q_{ij})$ with symmetrizing measure $(\pi_i > 0)$, we have density $J_{ij} = \pi_i q_{ij} \ (j \neq i)$ and $K_i = -\pi_i \sum_j q_{ij} \geq 0$ with respect to the counting measure.

Research supported in part by NSFC (No. 19631060), Math. Tian Yuan Found., Qiu Shi Sci. & Tech. Found., RFDP and MCME..

2. Liggett-Stroock form of Nash-type inequalities[17]

The main inequality we are interested in is the following:

$$\|f\|^2 \le CD(f)^{1/p}V(f)^{1/q}, \qquad f \in L^2(\pi), \tag{2}$$

where $\|f\| = \|f\|_2$, $C = C(p)$ is a constant and $1/p + 1/q = 1$ with $1 \le p < \infty$. Now, only the functional $V \ge 0$ has to be specified. The simplest case is that $p = 1$ and hence $1/q = 0$, then there is nothing to do about V and (2) is reduced to Poincaré inequality (1890):

$$\|f\|^2 \le CD(f), \qquad f \in L^2(\pi). \tag{3}$$

Thus, we may assume in what follows that $1 < p < \infty$. Next, since both $\|f\|^2$ and $D(f)$ have degree two of homogeneous, it is natural to assume that $V(cf) = c^2V(f)$ for all constant c. However, if we take $V(f) = D(f)$ or $\|f\|^2$, then (2) is again reduced to (3) for all $p \in (1, \infty)$.

Next, one may look at the other L_r-norm: $V(f) = \|f\|_r^2$, $r \ne 2$. We may assume that $r \in [1, 2)$ since the case of $r \in (2, \infty)$ can be reduced to the one of $r \in [1, 2)$ by the symmetry of the form. When $r = 1$, it is called Nash inequality (1958):

$$\|f\|^2 \le CD(f)^{1/p}\|f\|_1^{2/q}, \qquad f \in L^2(\pi). \tag{4}$$

By setting $p = 1 + 2/\nu$ ($\nu > 0$), one gets the more familiar form of Nash inequality $\|f\|^{2+4/\nu} \le CD(f)\|f\|_1^{4/\nu}$, $f \in L^2(\pi)$. Sometimes, the form $D(f)$ on the right-hand side of (4) is replaced by a new form $\widetilde{D}(f) = D(f) + \delta\|f\|^2$ for some $\delta \ge 0$. It is surprising but actually proved in [7] that for all $r \in [1, 2)$, inequality (2) with $V(f) = \|f\|_r^2$ ($r \in [1, 2)$) is equivalent to (4) and hence we need only to consider (4).

Of course, there are many other choices of V:

$$V(f) = \sup_x |f(x)|^2, \qquad \sup_{x \ne x_0}\left|\frac{f(x) - f(x_0)}{\rho(x, x_0)}\right|^2, \qquad \sup_{x \ne y}\left|\frac{f(y) - f(x)}{\rho(x, y)}\right|^2 = \mathrm{Lip}(f)^2,$$

where ρ is a distance and x_0 is a reference point in E. The last one was used by Liggett (1991) and we may call the corresponding inequality Liggett inequality.

3. Alternative form of (2)

From now on, we often restrict ourselves to the case that π is a probability measure and the form $(D, \mathcal{D}(D))$ satisfies $D(1) = 0$. Then, the right-hand side of (2) becomes zero for constant function $f = 1$. Thus, it is necessary to make a change of the left-hand side of (2). For this, one simply uses the variation of f: $\mathrm{Var}(f) = \pi(f^2) - \pi(f)^2$ instead of $\|f\|^2$, where $\pi(f) = \int f d\pi$. Then we obtain the alternative form of (2) as follows.

$$\text{L.S. inequality}: \qquad \mathrm{Var}(f) \le CD(f)^{1/p}V(f)^{1/q}, \qquad f \in L^2(\pi). \tag{5}$$

Certainly, this contains the alternative forms of the particular inequalities.

$$\text{Nash inequality}: \qquad \mathrm{Var}(f) \le CD(f)^{1/p}\|f\|_1^{2/q}.$$

$$\text{Poincaré inequality}: \qquad \mathrm{Var}(f) \le CD(f).$$

$$\text{Liggett inequality}: \qquad \mathrm{Var}(f) \le CD(f)^{1/p}\mathrm{Lip}(f)^{2/q}.$$

In what follows, unless otherwise stated, we mainly deal with these inequalities.

4. The second class of inequalities

Keeping the right-hand side of (3) but making a change of the left-hand side, one gets the following inequality[24]

$$\left\{ \int |f|^{2p/(p-1)} U\left(f^2/\|f\|^2\right) d\pi \right\}^{(p-1)/p} \leq C D(f), \qquad f \in L^2(\pi). \tag{6}$$

When $U = 1$, it is just *Sobolev inequality* (1936):

$$\|f\|_{2p/(p-1)}^2 \leq C D(f), \qquad f \in L^2(\pi).$$

Since $2p/(p-1) \geq 2$, the inequality is stronger than the Poincaré one unless $p = \infty$. When $U = \log$ and $p = \infty$, (6) is *logarithmic Sobolev inequality* (L. Gross, 1975):

$$\text{LogS}: \qquad \int f^2 \log\left(f^2/\|f\|^2\right) d\pi \leq C D(f), \qquad f \in L^2(\pi). \tag{7}$$

The advantage of the last inequality is that it is a powerful tool in the study of infinite-dimensional analysis but not the Sobolev one.

5. Importance of the inequalities

Denote by $(P_t)_{t \geq 0}$ the semigroup generated by the form $(D, \mathcal{D}(D))$. Then we have the following fundamental result.

Theorem 1[17] *Let $V : L^2(\pi) \to [0, \infty]$ satisfy $V(c_1 f + c_2) = c_1^2 V(f)$ for all constants $c_1, c_2 \in \mathbb{R}$.*

1. Assume additionally that $V(P_t f) \leq V(f)$ for all $t \geq 0$ and $f \in L^2(\pi)$ (it is automatic when $V(f) = \|f\|_r^2$). If (5) holds, then

$$\text{Var}(P_t f)\left(= \|P_t f - \pi(f)\|^2\right) \leq C V(f)/t^{q-1}, \qquad t > 0. \tag{8}$$

2. Conversely, (8) $\Longrightarrow$ (5).

Thus, inequality (5) describes L^2-algebraic convergence of the semigroup to its equilibrium state π. For Poincaré inequality, one indeed has L^2-exponential convergence:

$$\text{Var}(f) \leq C D(f) \iff \text{Var}(P_t f) \leq \text{Var}(f) e^{-2t/C}. \tag{9}$$

The smallest constant $C = \lambda_1^{-1}$, $\lambda_1 := \inf\{D(f) : \pi(f) = 0, \|f\| = 1\}$ is called the *spectral gap* of the form $(D, \mathcal{D}(D))$. The similar results hold for (2) and (3).

Based on (8) and (9), inequalities (5) and (7) now consist of the main tools in the study of phase transitions and the effectiveness of random algorithms.

6. Relation of the above inequalities

There is a simple comparison between the above inequalities:

$$\text{Nash ineq.} \Longrightarrow \text{LogS ineq.} \Longrightarrow \text{Poincaré ineq.} \Longrightarrow \text{Liggett ineq.} \tag{10}$$

Here "$\Longrightarrow$" means "implies" as usual but the last implication needs a mild condition. We will come back to this comparison in the last section. We remark that for (5) with $V(f) = \|f\|_r^2$, there is a jump at $r = 2$: For each $r < 2$, it is stronger than logarithmic Sobolev inequality but at $r = 2$, it becomes suddenly weaker than logarithmic Sobolev inequality.

7. Methods

As far as we know, there are two general and powerful methods to handle these inequalities.

a) The probabilistic method—coupling method. It has been successfully applied to the Riemannian geometry, elliptic operators and Markov chains. Refer to the survey articles [5] for the present status of the study and also for a comprehensive list of publications.

b) The second powerful method comes from Riemannian geometry, which is the one we are going to discuss here.

8. Isoperimetry

A very ancient geometric result says that among the different regions with fixed length of boundary, the circle has the largest area. That is, for a region A with smooth boundary ∂A, we have

$$\frac{|\partial A|}{|A|^{1/2}} \geq \frac{2\pi r}{\sqrt{\pi r^2}} = 2\sqrt{\pi}, \tag{11}$$

where $|A|$ denotes the volume (length, area) of A. The higher-dimensional analog is also true. That is the following *isoperimetric inequality*:

$$\frac{|\partial A|}{|A|^{(d-1)/d}} \geq \frac{|S_{d-1}|}{|B_d|^{(d-1)/d}}, \tag{12}$$

where B_d is the d-dimensional unit ball and S_{d-1} is its surface. The right-hand side is called the *isoperimetric constant*. Refer to [2] for more details.

9. Cheeger's constants

It is well known that isoperimetric inequality plays a critical role in the study of Sobolev-type inequality. In 1970, Cheeger observed that the same idea can also be used to study Poincaré inequality. To do so, Cheeger introduced the so-called *Cheeger's constants*:

$$h = \inf_{A \subset M} \frac{|\partial A|}{|A|}; \qquad k = \inf_{A \cup B = M} \frac{|\partial(A \cap B)|}{|A| \wedge |B|} \tag{13}$$

for a compact manifold M. Comparing (13) with (12), it follows that the power $(d-1)/d$ disappears here. Cheeger established the following *Cheeger inequalities*:

$$D(f) \geq \frac{h^2}{4} \|f\|^2; \qquad D(f) \geq \frac{k^2}{4} \mathrm{Var}(f). \tag{14}$$

For the first one, the Dirichlet boundary is imposed[3], [16], [25].

The Cheeger's technique was used to study the estimate of spectral gap for jump processes. For instance, it was proved by Lawler and Sokal (1988)[15] that $\lambda_1 \geq \dfrac{k^2}{2M}$, where $M = \sup_x q(x)$ and $k = \inf\limits_{\pi(A) \in (0,1)} \dfrac{\int_A \pi(\mathrm{d}x) q(x, A^c)}{\pi(A) \wedge \pi(A^c)}$. As far as we know, in the past ten years or so, this result has been collected into six books [1], [4], [10], [11], [21] and [22]. From the titles of the books, one sees a wider range of applications of the study on spectral gap. The main problem is that the lower estimate vanishes when one passes to the unbounded operators. The problem has been open for more than ten years.

For logarithmic Sobolev inequality, there is a large number of publications in the context of diffusion processes in the past two decades or more. However, there was no result in the context of jump processes until Diaconis and Saloff-Coste's paper appeared in 1996[12]. They proved that for finite Markov chains, the logarithmic constant $\sigma := \inf\left\{D(f)/\int f^2 \log[|f|/\|f\|] : \|f\| = 1\right\}$ satisfies $\sigma \geq \dfrac{2(1 - 2\pi_*)\lambda_1}{\log[1/\pi_* - 1]}$ under the condition that $\sum_j |q_{ij}| = 1$ for all i, where $\pi_* = \min_i \pi_i$. Clearly, for infinite E, $\pi_* = 0$ and so the result is meaningless. This is also a challenge open problem ([5]; Problem 13). For Nash inequality, we are in the same situation (cf. [21]).

10. Main theorem

We are now glad to be able to report some answers to the above open problems. To do so, we need some new types of isoperimetric or Cheeger's constants.

For Poincaré ineq.[9] $k^{(\alpha)} = \displaystyle\inf_{\pi(A)\in(0,1)} \dfrac{J^{(\alpha)}(A \times A^c)}{\pi(A) \wedge \pi(A^c)}$

For Nash ineq.[7] $k^{(\alpha)} = \displaystyle\inf_{\pi(A)\in(0,1)} \dfrac{J^{(\alpha)}(A \times A^c)}{[\pi(A) \wedge \pi(A^c)]^{(\nu-1)/\nu}}$, $\nu = 2(q-1)$

For LogS ineq.[24], [8] $k^{(\alpha)} = \displaystyle\lim_{r\to 0}\inf_{0<\pi(A)\leq r} \dfrac{J^{(\alpha)}(A \times A^c)}{\pi(A)\sqrt{\log[e + \pi(A)^{-1}]}}$

$k^{(\alpha)} = \displaystyle\lim_{\delta\to\infty}\inf_{\pi(A)>0} \dfrac{J^{(\alpha)}(A \times A^c) + \delta\pi(A)}{\pi(A)\sqrt{1 - \log\pi(A)}}$

Here only $J^{(\alpha)}$ has not defined yet. Note that the original kernel J can be very unbounded. To avoid this, choose a symmetric function $r(x, y)$ so that

$$J^{(1)}(\mathrm{d}x, E)/\pi(\mathrm{d}x) \leq 1, \qquad \pi\text{-a.e.,} \tag{15}$$

where $J^{(\alpha)}(\mathrm{d}x, \mathrm{d}y) = I_{\{r(x,y)>0\}}\dfrac{J(\mathrm{d}x, \mathrm{d}y)}{r(x, y)^{\alpha}}$ for $\alpha \in (0, 1]$ and $J^{(0)} = J$. For jump processes, one simply chooses $r(x, y) = q(x) \vee q(y)$. This key idea comes from [9]. Note that the second one is close to (12) and the first one is close to (13) since for manifolds, $|A| = \pi(A)$.

Having the constants at hand, it is a simple matter to state our main result.

Theorem 2 *Let π be a probability measure. For the form given by (1) with $K(\mathrm{d}x) = 0$, if $k^{(1/2)} > 0$, then the corresponding inequality in (5) or (7) holds.*

The theorem is proved in four papers [9], [7], [24] and [8] in which some explicit lower bounds in terms of $k^{(\alpha)}$ are also presented. We remark that even though the above condition $k^{(1/2)} > 0$ is in general not necessary but it can still be sharp qualitatively.

To give some impression about how the Cheeger's constants are related to the inequalities, we now sketch of the proof for the new type of Cheeger's inequalities which imply Poincaré ones.

11. Sketch of the proof

a) The first step is a simple observation about the set form and the functional form of the Cheeger's constants. Fix $B \in \mathcal{E}$ with $\pi(B) \in (0, 1)$. We have

$$h_B^{(\alpha)} = \inf_{A \subset B} \frac{J^{(\alpha)}(A \times A^c)}{\pi(A)} \quad \text{(set form)}$$

$$= \inf \left\{ \frac{1}{2} \int J^{(\alpha)}(\mathrm{d}x, \mathrm{d}y)|f(y) - f(x)| : f \geq 0, \, f|_{B^c} = 0, \, \|f\|_1 = 1 \right\}$$

$$\text{(functional form)}.$$

The proof is not hard. By taking $f = I_A$ ($A \subset B$), the previous one follows from the latter one. The other implication uses a co-area formula from geometry.

b) The next step is using Cauchy-Schwarz inequality. Let f satisfy $f|_{B^c} = 0$ and $\|f\| = 1$. From a) and condition (15), it follows that

$$h_B^{(1)^2} \leq \left\{ \frac{1}{2} \int J^{(1)}(\mathrm{d}x, \mathrm{d}y)|f(y)^2 - f(x)^2| \right\}^2$$

$$\leq \frac{1}{2} D^{(1)}(f) \int J^{(1)}(\mathrm{d}x, \mathrm{d}y)[f(y) + f(x)]^2$$

$$= \frac{1}{2} D^{(1)}(f) \int J^{(1)}(\mathrm{d}x, \mathrm{d}y) \left[2(f(y)^2 + f(x)^2) - (f(y) - f(x))^2 \right]$$

$$\leq D^{(1)}(f) \left[2 - D^{(1)}(f) \right].$$

This gives us $D^{(1)}(f) \geq 1 - \sqrt{1 - h_B^{(1)^2}}$.

c) By another use of Cauchy-Schwarz inequality and (15), we get $h_B^{(1/2)^2} \leq D(f)\left[2 - D^{(1)}(f) \right]$. Combining the above two estimates, we finally obtain a new type of the first Cheeger's inequality:

$$D(f) \geq \frac{h_B^{(1/2)^2}}{1 + \sqrt{1 - h_B^{(1)^2}}} = \frac{h_B^{(1/2)^2}}{1 + \sqrt{1 - h_B^{(1)^2}}} \|f\|^2, \qquad f|_{B^c} = 0.$$

From the proof, the relation between the constants $h_B^{(1/2)}$, $h_B^{(1)}$ and inequality (3) should be clear. Define $\lambda_0(B) = \inf\{D(f) : f|_{B^c} = 0, \|f\| = 1\}$. Then we have proved that

$$\lambda_0(B) \geq \frac{h_B^{(1/2)^2}}{1 + \sqrt{1 - h_B^{(1)^2}}}. \tag{16}$$

d) Finally, we adopt another idea due to Cheeger: the splitting technique. That is $\lambda_1 \geq \inf_{0 \leq \pi(B) \leq 1/2} \lambda_0(B)$ which is a key of the proof. Noting that $k^{(\alpha)} = \inf_{\pi(A) \in (0, 1/2]} h_B^{(\alpha)}$, one can deduce from (16) another new type of the second Cheeger's inequality (compare with (14)): $D(f) \geq \dfrac{k^{(1/2)^2}}{1 + \sqrt{1 - k^{(1)^2}}} \mathrm{Var}(f)$. That is,

$\lambda_1 \geq \dfrac{k^{(1/2)^2}}{1 + \sqrt{1 - k^{(1)^2}}}$. Refer to [9], [7], [24] and [8] for details and further references.

Finally, we discuss the relationship between the above inequalities and the traditional ergodic theory.

12. New story of ergodic theory

For simplicity, we restrict ourselves to continuous-time, irreducible Markov chains with transition probability matrix $P(t) = (p_{ij}(t))$ on a countable state space E. The chain is called *ergodic* if there exists a distribution $(\pi_i > 0 : i \in E)$ such that

$$\lim_{t \to \infty} p_{ij}(t) = \pi_j, \qquad i, j \in E. \tag{17}$$

It is called *exponentially ergodic* if there is a constant $\varepsilon > 0$ such that for all $i, j \in E$, there exists a constant C_{ij} so that

$$|p_{ij}(t) - \pi_j| \le C_{ij}e^{-\varepsilon t}, \qquad t > 0. \tag{18}$$

The chain is called *strongly ergodic* if

$$\lim_{t \to \infty} \sup_i |p_{ij}(t) - \pi_j| = 0. \tag{19}$$

These three types of ergodicity consist of the most common topics in the study of ergodic theory for Markov processes. It is known that strong ergodicity implies the uniformly exponential decay $\sup_i |p_{ij}(t) - \pi_j| \le C_j e^{-\varepsilon t}$ for some constants C_j and $\varepsilon > 0$. Hence

$$\text{Strong ergodicity} \Longrightarrow \text{Exponential ergodicity} \Longrightarrow \text{Ordinary ergodicity}. \tag{20}$$

A question arises naturally: Are there any relationship between the three types of ergodicity and the inequalities discussed above? The answer is affirmative, even though these two classes of objects look like very different.

Now, the new story of ergodic theory can be summarized as the following diagram, which is the main new result of this paper.

Theorem 3 *For reversible Markov chains, the following implications hold.*

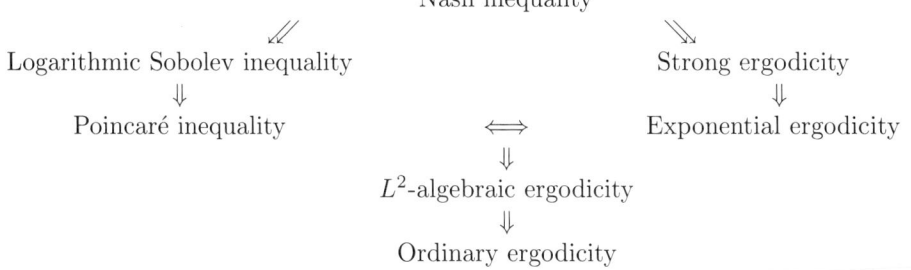

Here L^2-algebraic ergodicity means that (8) holds for some V satisfying the first assumption of Theorem 1 and $V(f) < \infty$ for all functions f with finite support.

Before moving on, let us make some remarks about the theorem. First, the most parts of the theorem also holds for general Markov processes under some mild assumption (see the proof given below for details). Next, the theorem is complete in the sense that all of the one-side implications "$\Longrightarrow$" can not be replaced by "$\Longleftrightarrow$". Besides, strong ergodicity and logarithmic Sobolev inequality are not comparable as will be shown soon. Thirdly, there are well known criteria for the three types of ergodicity in terms of Q-matrix, but none of them provides us any explicit "convergence rate". On the other hand, the study on the inequalities are often devoted to estimate the rates but up to now there is still no criterion for the inequalities in the publications. Thus, due to the equivalence given in the above theorem, the study on one side can benefit from the other. Finally, the inequalities are now powerful tools in infinite-dimensional situation, but the three types of ergodicity are more or less finite-dimensional objects. In any case, it is hoped that the diagram has made a meaningful change to the picture of ergodic theory.

The original purpose of the study on Liggett inequality is for L^2-algebraic convergence. However, the inequality depends heavily on the choice of the distance ρ. If

$$C_\rho := \int \pi(\mathrm{d}x)\pi(\mathrm{d}y)\rho(x,y)^2 < \infty,$$

then for every f with $\mathrm{Var}(f) = 1$, we have $1 = \frac{1}{2}\int \pi(\mathrm{d}x)\pi(\mathrm{d}y)[f(x) - f(y)]^2 \leq \frac{1}{2}C_\rho\mathrm{Lip}(f)^2$. Hence, replacing $V(f)$ with $\mathrm{Var}(f)$ on the right-hand side of (5), it follows that Poincaré inequality $\Longrightarrow$ Liggett inequality. This fact plus the assumptions of part (1) of Theorem 1 implies L^2-algebraic convergence with $V(f) =\mathrm{Lip}(f)^2$. Next, because $\mathrm{Lip}(I_{\{k\}}) = \sup_{j \neq k} \rho(k,j)^{-1}$, the last condition of Theorem 3 becomes $\inf_{j:j\neq k} \rho(k,j) > 0$ for all $k \in E$. Certainly, if one uses

$$V(f) = \sup_{x\neq x_0}\left|\frac{f(x) - f(x_0)}{\rho(x,x_0)}\right|^2 \quad \text{instead of } V(f) =\mathrm{Lip}(f)^2, \text{ the resulting conditions}$$

are different.

Proof Theorem 3 First, we prove the implication: "Nash inequality $\Longrightarrow$ Strong ergodicity". Assume that Nash inequality holds. Then by (8), we have $\|P_tf - \pi(f)\| \leq C\|f\|_1/t^{(q-1)/2}$ and so $\|(P_t - \pi)f\| \leq C\|f - \pi(f)\|_1/t^{(q-1)/2}$. This means that the operator norm $\|P_t - \pi\|_{1\to 2}$ as a mapping from $L^1(\pi)$ to $L^2(\pi)$ is bounded above by $C/t^{(q-1)/2}$. Because of the symmetry of $P_t - \pi$, we get $\|P_t - \pi\|_{1\to\infty} \leq \|P_t - \pi\|_{1\to 2}\|P_t - \pi\|_{2\to\infty} = \|P_t - \pi\|_{1\to 2}^2$. Hence $\sup_x \|P_t(x,\cdot) - \pi\|_{\mathrm{Var}} = \sup_x \sup_{|f|\leq 1}|(P_t(x,\cdot) - \pi)f| \leq \sup_x \sup_{\|f\|_1\leq 1}|(P_t(x,\cdot) - \pi)f| = \|P_t - \pi\|_{1\to\infty} \leq C/t^{q-1} \to 0$ as $t \to \infty$. This proves strong ergodicity of $(P_t(x,\mathrm{d}y))$.

The implications "Nash inequality $\Longrightarrow$ Logarithmic Sobolev inequality"and "Logarithmic Sobolev inequality $\Longrightarrow$ Poincaré inequality" are proved in [7] and [14] respectively. The implications "Poincaré inequality $\Longleftrightarrow$ L^2-exponential convergence $\Longrightarrow$ L^2-algebraic convergence" are obvious by (9). All the above proofs work for general reversible Markov processes. The implication "Exponential ergodicity $\Longleftrightarrow$ Poincaré inequality" is proved in [6] for Markov chains but it was mentioned there the result should work in more general setup. However, it is known that the equivalence fails in the infinite-dimensional situation, for instance when there exist phase transitions. To see that "L^2-algebraic convergence $\Longrightarrow$ Ordinary ergodicity", simply note that $\pi_i|p_{ik}(t) - \pi_k|^2 \leq \sum_j \pi_j|p_{jk}(t) - \pi_k|^2 \leq CV(I_{\{k\}})/t^{q-1} \to 0$ as $t \to \infty$.

Finally, one may replace "pointwise" by "total variation" in definition of the three types of ergodicity (these definitions are indeed equivalent to the ones given by (17)–(19) in the discrete case):

$$\text{Ordinary ergodicity :} \qquad \lim_{t\to\infty} \|p_t(x,\cdot) - \pi\|_{\mathrm{Var}} = 0$$

$$\text{Exponential ergodicity :} \qquad \|p_t(x,\cdot) - \pi\|_{\mathrm{Var}} \leq C(x)e^{-\varepsilon t}$$

$$\text{Strong ergodicity :} \qquad \lim_{t\to\infty} \sup_x \|p_t(x,\cdot) - \pi\|_{\mathrm{Var}} = 0.$$

Then, the implications in (20) hold for general Markov processes when $\mathbb{E}$ is countably generated. In other words, if the Markov process corresponding to the semigroup (P_t) is irreducible and aperiodic in the Harris sense, then (20) holds. To see this, noting that by [4; §4.4] and [13], the continuous-time case can be reduced to the discrete-time one and then the conclusion follows from [18; Chapter 16]. □

We now show by some examples that strong ergodicity and logarithmic Sobolev inequality are not comparable. To do so, we need the following result taken from [26] and [27].

Theorem 4 *For regular birth-death process with birth rates b_i $(i \geq 0)$ and death rates a_i $(i \geq 1)$, the process is strong ergodic iff $S := \sum_{n=1}^{\infty} a_{n+1}^{-1} \{1 + \sum_{k=1}^{n} b_k \cdots b_n / a_k \cdots a_n\} < \infty$.*

Example 5 *Consider the birth-death process with either (1) $b_i = a_i = i^2 \log^\gamma i$ $(i \geq 1)$, $\gamma \in \mathbb{R}$ or (2) $b_i = i^\gamma / 2$ and $a_i = i^\gamma$ $(i \geq 1)$, $\gamma \geq 0$ and $b_0 = 1$. Then logarithmic Sobolev inequality holds iff $\gamma \geq 1$ and the process is strongly ergodic iff $\gamma > 1$.*

Proof The second assertion follows from Theorem 4. The first assertion is proved in [24] and [7; Examples 2.6 and 2.9]. The key idea is that for logarithmic Sobolev inequality of birth-death process, the constant $k^{(1/2)}$ used in Theorem 2 is equal to

$$\xi := \inf_{i \geq 1} \pi_i a_i / \left(\sum_{j \geq i} \pi_j\right) \sqrt{r_{i,i-1}\left(1 - \log \sum_{j \geq i} \pi_j\right)} > 0,$$

where $r_{ij} = (a_i + b_i) \vee (a_j + b_j)$. Noting that $\pi_i a_i = \pi_{i-1} b_{i-1}$, we have $\xi > 0$ iff $\sup_{k \geq 1} \frac{1}{\pi_k b_k} \left(\sum_{j \geq k+1} \pi_j\right) \left[r_{k+1,k}\left(1 - \log \sum_{j \geq k+1} \pi_j\right)\right]^{1/2} < \infty$. On the other hand, it is easy to confirm that $S < \infty$ iff $\sum_{k \geq 0} \frac{1}{\pi_k b_k} \sum_{j \geq k+1} \pi_j < \infty$. These facts indicate the relationship between "$S < \infty$" and "$\xi > 0$". □

Example 6 Given a distribution $(\pi_i > 0)$ on a countable set E. Let $q_{ij} = \pi_j$ for all $j \neq i$ and $q_i = -q_{ii} = 1 - \pi_i$. Then the process is strongly ergodic but logarithmic Sobolev inequality fails.

Proof It is proved in [24] that logarithmic Sobolev inequality does not hold for bounded operators in infinite space and so the second assertion follows. The proof is simply applying (7) to $f = I_A / \sqrt{\pi(A)}$ with $\pi(A) \in (0, 1)$ and then letting $\pi(A) \to 0$.

To prove the first assertion, for simplicity, let $0 \in E$. Next, let $f_0 = 0$ and $f_i = 1/\pi_0$ $(i \neq 0)$. Then $\pi(f) := \sum_j \pi_j f_j = \pi_0^{-1} - 1$. Hence $\sum_{j \neq i} q_{ij}(f_j - f_i) = \pi(f) - f_i = -1$ for all $i \neq 0$ and $\sum_{j \neq 0} q_{0j}(f_j - f_0) = \pi(f) < \infty$. Because $0 \leq f \leq \pi_0^{-1}$, the first assertion now follows from the well know criterion for strong ergodicity (cf. [4; Part (4) of Theorem 4.45]). □

References

[1] Aldous, D. G. & Fill, J. A. (1994–), *Reversible Markov Chains and Random Walks on Graphs*, URL **www. stat.Berkeley.edu/users/aldous/book.html** .

[2] Berger, E. and Gostiaus, B. (1988) *Differential Geometry: Manifolds, Curves, and Surfaces*

[3] Cheeger, J. (1970) *A lower bound for the smallest eigenvalue of the Laplacian* Problems in analysis, a symposium in honor of S. Bochner 195–199, Princeton U. Press, Princeton.

[4] Chen, M. F. (1992) *From Markov Chains to Non-Equilibrium Particle Systems*, World Scientific .

[5] Chen, M. F. (1997) *Coupling, spectral gap and related topics,* (I): Chin. Sci. Bulletin, 42:14, 1472–1477 (Chinese Edition); 42:16, 1321–1327 (English Edition). (II): 42:15, 1585–1591 (Chinese Edition); 42:17, 1409–1416 (English Edition). (III): 42:16, 1696–1703 (Chinese Edition); 1997, 42:18, 1497–1505 (English Edition) .

[6] Chen, M. F. (1998) *Equivalence of exponential ergodicity and L^2-exponential convergence for Markov chains* preprint.

[7] 7 Chen, M. F. (1999) *Nash inequalities for general symmetric forms* to appear in Acta Math. Sin. Eng. Ser.

[8] 8 Chen, M. F. (1999) *Logarithmic Sobolev inequality for symmetric forms* preprint.

[9] 9 Chen, M. F. and Wang, F. Y. (1998) *Cheeger's inequalities for general symmetric forms and existence criteria for spectral gap* Preprint. Abstract. Chin. Sci. Bulletin 43:14, 1475–1477 (Chinese Edition); 43:18, 1516–1519 (English Edition) .

[10] 10 Chung, F. R. K. (1997) *Spectral Graph Theory CBMS,* **92**, AMS, Providence, Rhode Island .

[11] 11 Colin de Verdière, Y. (1998) *Spectres de Graphes Soc. Math. France.*

[12] 12 Diaconis, P. and Saloff-Coste, L. *Logarithmic Sobolev incqualities for finite Markov chains* Ann. Appl. Prob. 1996, 695–750.

[13] 13 Down, D., Meyn, S. P. and Tweedie, R. L. (1995) *Exponential and uniform ergodicity of Markov processes* Ann. Prob. 23, 1671–1691.

[14] 14 Gross, L. (1976) *Logarithmic Sobolev inequalities* Amer. J. Math. 97, 1061–1083 .

[15] 15 Lawler, G. F. and Sokal, A. D. (1988) *Bounds on the L^2 spectrum for Markov chain and Markov processes: a generalization of Cheeger's inequality* Trans. Amer. Math. Soc. **309**, 557–580.

[16] 16 Li, P. (1993) *Lecture Notes on Geometric Analysis Seoul National Univ., Korea.*

[17] 17 Liggett, T. M. (1991) *L_2 rates of convergence for attractive reversible nearest particle systems: the critical case* Ann. of Prob. 19:3, 935–959 .

[18] 18 Meyn, S. P. & Tweedie, R. L. (1993) *Markov chains and Stochastic Stability Springer-Verlag .*

[19] 19 Nash, J. (1958) *Continuity of solutions of parabolic and elliptic equations* Amer. J. Math. 80, 931–954.

[20] 20 Poincaré, H. (1890) *Sur les équations aux dérivées partielles de la physique mathématique* Amer. J. Math. 12, 211–294 .

[21] 21 Saloff-Coste, L. (1997) *Lectures on finite Markov chains* LNM **1665**, 301–413, Springer-Verlag.

[22] 22 Sinclair, A. (1993) *Algorithms for Random Generation and Counting: A Markov Chain Approach Birk-häuser.*

[23] 23 Sobolev, S. (1936) *On a theorem in functional analysis* Mat. Sbornik 4 (1938), 471–497 (In Russian); AMS Translations, Ser. 2, 34 (1963), 39–68 .

[24] 24 Wang, F. Y. (1998) *Sobolev type inequalities for general symmetric forms* to appear in Proc. of Amer. Math. Soc.

[25] 25 Yau, S. T., Schoen, R. (1988) *Differential Geometry Science Press* (In Chinese)*, Beijing, China.*

[26] 26 Zhang, Y. H. (1999) *Strong ergodicity for continuous-time Markov chains* preprint .

[27] 27 Zhang, H. J., Lin, X. and Hou, Z. T. (1998) *Uniformly polynomial convergence for standard transition functions preprint.*

DEPARTMENT OF MATHEMATICS, BEIJING NORMAL UNIVERSITY, BEIJING 100875, THE PEOPLE'S REPUBLIC OF CHINA. E-MAIL: MFCHENBNU.EDU.CN

AMS/IP Studies in Advanced Mathematics
Volume 26, 2002

Solvability of a Stochastic Linear Quadratic Optimal Control Problem

Shuping Chen[1] and Jiongmin Yong[2]

ABSTRACT. This paper discusses the solvability of a stochastic linear quadratic optimal control problem (LQ problem, for short) which allowing the cost functional to have a negative weight on the square of the control variable. By the recently developed results on linear forward-backward stochastic differential equations (FBSDEs, for short), we establish some sufficient conditions for the solvability of the LQ problem, and the associated Riccati equation.

1. Introduction

Let $(\Omega, \mathcal{F}, \mathbf{P}, \{\mathcal{F}_t\}_{t\geq 0})$ be a complete filtered probability space on which a one-dimen-
sional standard Brownian motion $w(\cdot)$ is defined such that $\{\mathcal{F}_t\}_{t\geq 0}$ is the natural filtration generated by $w(\cdot)$, augmented by all the $\mathbf{P}$-null sets in $\mathcal{F}$. We consider the following linear controlled stochastic differential equation:

$$(1.1) \quad \begin{cases} dx(t) = [Ax(t) + Bu(t)]dt + [Cx(t) + Du(t)]dw(t), & t \in [s, T], \\ x(s) = \xi, \end{cases}$$

where A, B, C, D are matrices of suitable dimensions, $s \in [0, T)$, and $\xi \in \mathbb{R}^n$. In the above, $u(\cdot)$ is a *control process* and $x(\cdot)$ is the corresponding state process. Let $\mathcal{U}[s, T] = L_{\mathcal{F}}^2(s, T; \mathbb{R}^m)$, the set of all $\mathbb{R}^m$-valued $\{\mathcal{F}_t\}_{t\geq 0}$-adapted square-integrable processes defined on $[s, T]$. The control process $u(\cdot)$ is taken from $\mathcal{U}[s, T]$.

Clearly, for any $(\xi, u(\cdot)) \in \mathbb{R}^n \times \mathcal{U}[s, T]$, there exists a unique (strong) solution $x(\cdot) \in L_{\mathcal{F}}^2(s, T; \mathbb{R}^n)$ to (1.1). Thus, we can define a *cost functional* as follows:

$$(1.2) \qquad J(s, \xi; u(\cdot)) =$$
$$E\left\{ \int_s^T [\langle Qx(t), x(t) \rangle + \langle Ru(t), u(t) \rangle] dt + \langle Gx(T), x(T) \rangle \right\},$$

1991 *Mathematics Subject Classification.* 93E20, 49K45, 49N10, 60H10.

Key words and phrases. Stochastic LQ problem, FBSDEs, Riccati equation.

This author is supported in part by the NSFC and the Chinese State Education Ministry Science Foundation.

This author is supported in part by the NSFC, under grant 79790130, the National Distinguished Youth Science Foundation of China under grant 19725106, the Chinese Education Ministry Science Foundation under grant 97024607, and the Li Foundation.

where Q, R and G are symmetric matrices of proper dimensions. Let us now state the stochastic linear quadratic optimal control problem as follows:

Problem (LQ). For each $s \in [0, T)$ and $\xi \in \mathbb{R}^n$, find a $\bar{u}(\cdot) \in \mathcal{U}[s, T]$ such that

$$
(1.3) \qquad J(s, \xi; \bar{u}(\cdot)) = \inf_{u(\cdot) \in \mathcal{U}[s,T]} J(s, \xi; u(\cdot)) \stackrel{\Delta}{=} V(s, \xi).
$$

We call $V(\cdot, \cdot)$ the *value function* of Problem (LQ). Note that

$$
(1.4) \qquad V(T, \xi) = \langle G\xi, \xi \rangle, \qquad \forall \xi \in \mathbb{R}^n.
$$

We point out here that we do not assume the positive (semi-)definite for the matrices Q, R, G. Such kind of stochastic LQ problems have been studied recently (see [3–6], and also [15]). For some earlier literature, we refer the readers to [13,1,9,11], etc., and the reference list given in [15].

In this paper, we confine ourselves within the case that all the coefficients are (deterministic) constant matrices. The main purpose of the paper is to present some sufficient conditions for the solvability of Problem (LQ), under certain conditions. The main ingredients are the results developed in [14] for forward-backward stochastic differential equations. At the same time, the solvability of the corresponding Riccati equation will also be presented.

The rest of the paper is organized as follows. In Section 2, we present some preliminary results, mainly recall some results obtained in [4,5]. Section 3 is devoted to the solvability of Problem (LQ) and the Riccati equation via results of linear FBSDEs. In Section 4, we present a nontrivial example to illustrate the results developed in this paper.

2. Preliminaries

Let $\mathcal{S}^n$ be the set of all $n \times n$ symmetric matrices. For any Euclidean space X, let $L_{\mathcal{F}}^\infty(0, T; X)$ (resp. $C_{\mathcal{F}}([0, T]; X)$) be the set of all X-valued $\{\mathcal{F}_t\}_{t \geq 0}$-adapted bounded (resp. bounded continuous) processes, and $L_{\mathcal{F}_T}^\infty(\Omega; X)$ be the set of all X-valued $\mathcal{F}_T$-measurable bound-ed random variables. Here, X could be $\mathbb{R}^n$, $\mathbb{R}^{n \times n}$, $\mathcal{S}^n$, etc.

Let us recall the following definitions ([4,5]).

Definition 2.1. Problem (LQ) is said to be

(i) finite at $(s, \xi) \in [0, T] \times \mathbb{R}^n$ if the right hand side of (1.3) is finite.

(ii) *(uniquely) solvable* at $(s, \xi) \in [0, T] \times \mathbb{R}^n$ if there exists a (unique) control $\bar{u}(\cdot) \in \mathcal{U}[s, T]$ such that (1.3) holds. In this case, $\bar{u}(\cdot)$ is called an *optimal control*, the corresponding $\bar{x}(\cdot)$ is called an *optimal state process*, and $(\bar{x}(\cdot), \bar{u}(\cdot))$ is called an *optimal pair*.

(iii) If for $s \in [0, T]$, Problem (LQ) is finite (resp. (uniquely) solvable) at all (s, ξ), we say that Problem (LQ) is *finite* (resp. (*uniquely*) *solvable*) at s. If Problem (LQ) is finite (resp. (uniquely) solvable) at all $s \in [0, T]$, we say that Problem (LQ) is *finite* (resp. (*uniquely*) *solvable*).

Next, let us introduce the following stochastic differential equation:

$$(2.1) \quad \begin{cases} dx(t) = [Ax(t) + Bu(t)]dt + [Cx(t) + Du(t)]dw(t), \\ dy(t) = [Qx(t) - A^T y(t) - C^T z(t)]dt + z(t)dw(t), \\ x(s) = \xi, \qquad y(T) = -Gx(T). \end{cases}$$

Such a system is called a *forward-backward stochastic differential equations* (FB-SDE, for short) since the equation for $x(\cdot)$ is *forward* (meaning that it is an initial value problem, which is to be solved forward) and the equation for $y(\cdot)$ (and $z(\cdot)$) is *backward* (meaning that it is a terminal value problem, which is to be solved backward). For given $(\xi, u(\cdot)) \in \mathbb{R}^n \times \mathcal{U}[s, T]$, an *adapted solution* of (2.1) is a triple of $\{\mathcal{F}_t\}_{t \geq 0}$-adapted processes $(x(\cdot), y(\cdot), z(\cdot))$ satisfying (2.1). It is clear that for any given $(\xi, u(\cdot)) \in \mathbb{R}^n \times \mathcal{U}[s, T]$, we can first solve the forward equation for $x(\cdot)$ and then solve (see [12]) She backward equation for $(y(\cdot), z(\cdot))$. Hence, for any $(\xi, u(\cdot)) \in \mathbb{R}^n \times \mathcal{U}[s, T]$, there exists a unique adapted solution $(x(\cdot), y(\cdot), z(\cdot))$ to (2.1). See [14,10] for more information concerning general theory of FBSDEs. (see [2,12], also.)

Let us also introduce the following Riccati equation associated with Problem (LQ).
(2.2)
$$\begin{cases} \dot{P}(t) = -P(t)A - A^T P(t) - Q \\ \qquad + [P(t)B + C^T P(t)D][R + D^T P(t)D]^{-1}[B^T P(t) + D^T P(t)C], \\ \qquad t \in [s, T], \\ P(T) = G, \\ R + D^T P(t)D > 0, \qquad t \in [s, T]. \end{cases}$$

The following theorem collects the results obtained in [4,5].

Theorem 2.2. (i) *Suppose Problem (LQ) is finite at some $(s, \xi) \in [0, T) \times \mathbb{R}^n$. Then*

$$(2.3) \qquad R + D^T GD \geq 0, \qquad \text{a.s.}$$

(ii) *Problem (LQ) is (uniquely) solvable at $(s, \xi) \in [0, T] \times \mathbb{R}^n$ with an (the) optimal pair $(\bar{x}(\cdot), \bar{u}(\cdot))$ if and only if Problem (LQ) is finite at (s, ξ), and the FBSDE (2.1) admits a (unique) adapted solution $(\bar{x}(\cdot), \bar{y}(\cdot), \bar{z}(\cdot))$ such that*

$$(2.4) \qquad R\bar{u}(t) + B^T \bar{y}(t) + D^T \bar{z}(t) = 0, \quad \text{a.e. } t \in [s, T], \text{ a.s.}$$

In addition, if R^{-1} exists, then an (the) optimal control $\bar{u}(\cdot)$ admits a representation:

$$(2.5) \qquad \bar{u}(t) = -R^{-1}[B^T \bar{y}(t) + C^T \bar{z}(t)], \quad t \in [\tau, T].$$

(iii) *Suppose for $s \in [0, T)$, (2.2) admits a solution $P(\cdot) : [s, T] \to \mathcal{S}^n$. Then Problem (LQ) is solvable at any $r \in [s, T]$ with the optimal control $\bar{u}(\cdot)$ being of state feedback form:*

$$(2.6) \qquad \bar{u}(t) = -[R + D^T P(t)D]^{-1}[B^T P(t) + D^T P(t)C]x(t), \qquad t \in [r, T],$$

and with the value function represented by

$$(2.7) \qquad V(r, \xi) = \langle P(r)\xi, \xi \rangle, \quad \forall (r, \xi) \in [s, T] \times \mathbb{R}^n.$$

We note that the above results are incomplete in the sense that we still do not know when Problem (LQ) is solvable, unless we know when either the FBSDE (2.1) or the Riccati equation (2.2) is solvable. Our purpose is to present some sufficient conditions for the solvability of (2.1) and/or (2.2).

3. Solvability of Problem (LQ).

We now introduce the following assumption.

(H) Matrix R has an inverse. Moreover,

(3.1) $$BR^{-1}D^T = 0, \quad C = 0,$$

and

(3.2) $$R + D^T G D > 0.$$

We emphasize that R is not necessary positive semi-definite. Also, B and D are not necessarily zero. Here is an simple example.

(3.3) $$B = D = \begin{pmatrix} 1 & 1 \\ 1 & 1 \end{pmatrix}, \quad R - \begin{pmatrix} 1 & 0 \\ 0 & -1 \end{pmatrix}$$

Now, we assume (H). Let $(x(\cdot), u(\cdot))$ be an optimal pair of Problem (LQ) and $(y(\cdot), z(\cdot))$ be the adapted solution of BSDE in (2.1). Then, plugging (2.5) into (2.1), we obtain

(3.4) $$\begin{cases} dx(t) = \big[Ax(t) + BR^{-1}B^T y(t)\big]dt + DR^{-1}D^T z(t)dw(t), \\ dy(t) = \big[Qx(t) - A^T y(t)\big]dt + z(t)dw(t), \\ x(s) = \xi, \quad y(T) = -Gx(T). \end{cases}$$

This is a coupled linear FBSDE. To obtain sufficient conditions for solvability of such an FBSDE via the result of [14], let us make some reductions. For notational simplicity, in what follows, we suppress the argument t. Let $\eta = y + Gx$. Then

(3.5)
$$\begin{aligned} d\eta &= dy + Gdx \\ &= \big[Qx - A^T y + G(Ax + BR^{-1}B^T y)\big]dt + \big[I + GDR^{-1}D^T\big]zdw \\ &= \big[(Q + GA)x + (-A^T + GBR^{-1}B^T)(\eta - Gx)\big]dt \\ &\qquad + (I + GDR^{-1}D^T)zdw \\ &= \big[(Q + GA + A^T G - GBR^{-1}B^T G)x + (-A^T + GBR^{-1}B^T)\eta\big]dt \\ &\qquad + (I + GDR^{-1}D^T)zdw. \end{aligned}$$

We define

(3.6) $$\zeta = (I + GDR^{-1}D^T)z.$$

A direct computation shows that

(3.7) $$(I + GDR^{-1}D^T)^{-1} = I - GD(R + D^T GD)^{-1}D^T.$$

This means that $(I + GDR^{-1}D^T)^{-1}$ exists and from (3.6), we have

(3.8) $$z = (I + GDR^{-1}D^T)^{-1}\zeta.$$

Consequently,

(3.9)
$$\begin{aligned} dx &= \big[Ax + BR^{-1}B^T(\eta - Gx)\big]dt + DR^{-1}D^T(I + GDR^{-1}D^T)^{-1}\zeta dw \\ &= \big[(A - BR^{-1}B^T G)x + BR^{-1}B^T\eta\big]dt \\ &\qquad + DR^{-1}D^T(I + GDR^{-1}D^T)^{-1}\zeta dw. \end{aligned}$$

Also,

(3.10) $$d\eta = \big[(Q + GA + A^T G - GBR^{-1}B^T G)x + (-A^T + GBR^{-1}B^T)\eta\big]dt + \zeta dw.$$

Hence, (3.4) becomes:

(3.11)
$$\begin{cases} dx(t) = \big[(A - BR^{-1}B^T G)x(t) + BR^{-1}B^T \eta(t)\big]dt \\ \qquad\qquad +DR^{-1}D^T(I + GDR^{-1}D^T)^{-1}\zeta(t)dw(t), \\ d\eta(t) = \big[(Q + GA + A^T G - GBR^{-1}B^T G)x(t) + (-A^T + GBR^{-1}B^T)\eta(t)\big]dt \\ \qquad\qquad +\zeta(t)dw(t), \\ x(s) = \xi, \qquad \eta(T) = 0. \end{cases}$$

Similar to [Y], we denote

$$\mathcal{A} \triangleq \begin{pmatrix} A - BR^{-1}B^T G & BR^{-1}B^T \\ Q + GA + A^T G - GBR^{-1}B^T G & -A^T + GBR^{-1}B^T \end{pmatrix},$$

(3.12)
$$\mathcal{C}_1 \triangleq \begin{pmatrix} D(R + D^T GD)^{-1}D^T \\ I \end{pmatrix}.$$

Then (3.11) can further be written as

(3.13)
$$\begin{cases} d\begin{pmatrix} x(t) \\ \eta(t) \end{pmatrix} = \mathcal{A}\begin{pmatrix} x(t) \\ \eta(t) \end{pmatrix} dt + \mathcal{C}_1\zeta(t)dw(t), \\ x(s) = \xi, \quad \eta(T) = 0. \end{cases}$$

Using variation of constants formula, we see that if $(x(\cdot), \eta(\cdot), \zeta(\cdot))$ is an adapted solution of (3.13), one must have some $\eta_0 \in \mathbb{R}^n$ such that

(3.14)
$$\begin{pmatrix} x(t) \\ \eta(t) \end{pmatrix} = e^{\mathcal{A}(t-s)}\begin{pmatrix} \xi \\ \eta_0 \end{pmatrix} + \int_s^t e^{\mathcal{A}(t-r)}\mathcal{C}_1\zeta(r)dw(r), \quad t \in [s, T],$$

and

(3.15)
$$\begin{aligned} 0 &= \begin{pmatrix} 0 & I \end{pmatrix}\begin{pmatrix} x(T) \\ \eta(T) \end{pmatrix} \\ &= \begin{pmatrix} 0 & I \end{pmatrix}\left\{ e^{\mathcal{A}(T-s)}\begin{pmatrix} \xi \\ \eta_0 \end{pmatrix} + \int_s^T e^{\mathcal{A}(T-r)}\mathcal{C}_1\zeta(r)dw(r)\right\} \\ &= \begin{pmatrix} 0 & I \end{pmatrix} e^{\mathcal{A}(T-s)}\begin{pmatrix} I \\ 0 \end{pmatrix}\xi + \begin{pmatrix} 0 & I \end{pmatrix} e^{\mathcal{A}(T-s)}\begin{pmatrix} 0 \\ I \end{pmatrix}\eta_0 \\ &\quad + \begin{pmatrix} 0 & I \end{pmatrix}\int_s^T e^{\mathcal{A}(T-r)}\mathcal{C}_1\zeta(r)dw(r), \end{aligned}$$

which is equivalent to the following:

(3.16)
$$\begin{aligned} \eta_1 &\triangleq -\begin{pmatrix} 0 & I \end{pmatrix} e^{\mathcal{A}(T-s)}\begin{pmatrix} I \\ 0 \end{pmatrix}\xi \\ &= \begin{pmatrix} 0 & I \end{pmatrix} e^{\mathcal{A}(T-s)}\begin{pmatrix} 0 \\ I \end{pmatrix}\eta_0 + \begin{pmatrix} 0 & I \end{pmatrix}\int_s^T e^{\mathcal{A}(T-r)}\mathcal{C}_1\zeta(r)dw(r). \end{aligned}$$

Hence, we see that (3.13) is solvable for any $\xi \in \mathbb{R}^n$ if and only if for any η_1 of form given in (3.16), the following is solvable:

(3.17)
$$\begin{cases} d\begin{pmatrix} x(t) \\ \eta(t) \end{pmatrix} = \mathcal{A}\begin{pmatrix} x(t) \\ \eta(t) \end{pmatrix} dt + \mathcal{C}_1\zeta(t)dw(t), \\ x(s) = 0, \quad \eta(T) = \eta_1. \end{cases}$$

Now, at this moment, the results of [14] (with a proper modification) applies. More precisely, (3.17) is solvable if

(3.18)
$$\begin{cases} \det\left\{ \begin{pmatrix} 0 & I \end{pmatrix} e^{\mathcal{A}(T-s)} \begin{pmatrix} 0 \\ I \end{pmatrix} \right\} \neq 0, \\ \det\left\{ \begin{pmatrix} 0 & I \end{pmatrix} e^{\mathcal{A}(t-s)} \mathcal{C}_1 \right\} > 0, \quad t \in [s, T]. \end{cases}$$

Hence, we obtain the following results for Problem (LQ).

Theorem 3.1. *Let (H) hold. Let (3.18) hold for some $s \in [0, T)$. Then, for any $\xi \in \mathbb{R}^n$, FBSDE (2.1) admits a unique adapted solution $(x(\cdot), y(\cdot), z(\cdot))$, which implies that Problem (LQ) is uniquely solvable at s.*

Next, let us look at the solvability of Riccati equation (2.2). Under assumption, (2.2) reads
(3.19)
$$\begin{cases} \dot{P}(t) = -P(t)A - A^T P(t) - Q + P(t)B[R + D^T P(t)D]^{-1}B^T P(t), \ t \in [s, T], \\ P(T) = G, \\ R + D^T P(t)D > 0, \qquad t \in [s, T]. \end{cases}$$

The following lemma is interesting.

Lemma 3.2. *Let (H) hold. Then $P(\cdot)$ is a solution of (3.19) if and only if it is a solution of the following:*

(3.20)
$$\begin{cases} \dot{P}(t) = -P(t)A - A^T P(t) - Q + P(t)BR^{-1}B^T P(t), \ t \in [s, T], \\ P(T) = G, \\ R + D^T P(t)D > 0, \qquad t \in [s, T] \end{cases}.$$

Proof. Suppose $P(\cdot)$ is a solution of (3.19). Then (we suppress t, for simplicity)
(3.21)
$$\begin{aligned} B(R + D^T PD)^{-1} - BR^{-1} &= B[(R + D^T PD)^{-1} - R^{-1}] \\ &= BR^{-1}[R - R - D^T PD](R + D^T PD)^{-1} \\ &= -BR^{-1}D^T PD(R + D^T PD)^{-1} = 0. \end{aligned}$$

Thus, $P(\cdot)$ is a solution of (3.20). Conversely, if $P(\cdot)$ is a solution of (3.20), then (3.21) holds and hence $P(\cdot)$ is a solution of (3.19). $\qquad\square$

We note that without looking at the third constraint, (3.20) looks like a standard Riccati equation for deterministic LQ problems. However, we should remind the readers that there is no positive semi-definiteness of either Q and R. Also, the third constraint is not obviously to be satisfied.

Next, by [14], if $(x(\cdot), \eta(\cdot), \zeta(\cdot))$ is an adapted solution of FBSDE (3.17) and the we hope to have relation

(3.22)
$$\eta(t) = \Pi(t)x(t), \qquad t \in [0, T],$$

then $\Pi(\cdot)$ should satisfy the following Riccati equation:

(3.23)
$$\begin{cases} \dot{\Pi}(t) + \Pi(t)(A - BR^{-1}B^T G) + (A - BR^1 B^T G)^T \Pi(t) \\ \qquad + \Pi(t)BR^{-1}B^T \Pi(t) - [Q + GA + A^T G - GBR^{-1}B^T G] = 0, \\ \Pi(T) = 0. \end{cases}$$

The following was proved in [14].

Theorem 3.3. *Let (H) hold and for some $s \in [0, T)$,*

$$(3.24) \quad \begin{cases} \det\left\{ \begin{pmatrix} 0 & I \end{pmatrix} e^{\mathcal{A}(t-s)} \begin{pmatrix} 0 \\ I \end{pmatrix} \right\} > 0, \\ \det\left\{ \begin{pmatrix} 0 & I \end{pmatrix} e^{\mathcal{A}(t-s)} \mathcal{C}_1 \right\} > 0, \\ t \in [s, T]. \end{cases}$$

Then (3.23) admits a unique solution $\Pi(\cdot)$ which has the following representation

$$(3.25) \quad \Pi(t) = -\left[\begin{pmatrix} 0 & I \end{pmatrix} e^{\mathcal{A}(T-t)} \begin{pmatrix} 0 \\ I \end{pmatrix} \right]^{-1} \begin{pmatrix} 0 & I \end{pmatrix} e^{\mathcal{A}(T-t)} \begin{pmatrix} I \\ 0 \end{pmatrix}, \quad t \in [s, T].$$

Consequently, FBSDE (3.17) admits a unique adapted solution $(x(\cdot), \eta(\cdot), \zeta(\cdot))$.

The following result reveals the relation between (3.20) and (3.23).

Proposition 3.4. *Let (H) hold. Let $P(\cdot)$ and $\Pi(\cdot)$ be solutions of (3.20) and (3.23), respectively. Then the following holds*

$$(3.26) \quad P(t) = G - \Pi(t), \quad t \in [s, T].$$

The proof is straightforward.

Combining Theorems 2.2 and 3.3, and Proposition 3.4, we see that when (H) and (3.24) hold, Riccati equation (3.20) admits a unique solution $P(\cdot)$, which leads to the solvability of Problem (LQ). It is important to note that conditions (H) and (3.24) are checkable, in principle.

4. An Example.

In this section, we present an example for which our theory applies.

Example 4.1. Let

$$(4.1) \quad \begin{cases} A = \begin{pmatrix} 0 & 1 \\ 0 & 0 \end{pmatrix}, \ B = \begin{pmatrix} 1 & 0 \\ 0 & 0 \end{pmatrix}, \ C = 0, \ D = \begin{pmatrix} 0 & 0 \\ 0 & 1 \end{pmatrix}, \\ Q = \begin{pmatrix} 1 & 1 \\ 1 & 0 \end{pmatrix}, \ R = \begin{pmatrix} 1 & 0 \\ 0 & -1 \end{pmatrix}, \ G = \begin{pmatrix} -1 & 0 \\ 0 & 2 \end{pmatrix}. \end{cases}$$

Then we have

$$(4.2) \quad \begin{aligned} BR^{-1}D^T &= \begin{pmatrix} 1 & 0 \\ 0 & 0 \end{pmatrix} \begin{pmatrix} 1 & 0 \\ 0 & -1 \end{pmatrix} \begin{pmatrix} 0 & 0 \\ 0 & 1 \end{pmatrix} = 0, \\ R + D^T G D &= \begin{pmatrix} 1 & 0 \\ 0 & -1 \end{pmatrix} + \begin{pmatrix} 0 & 0 \\ 0 & 1 \end{pmatrix} \begin{pmatrix} 0 0 - 1 & 0 \\ 0 & 2 \end{pmatrix} \begin{pmatrix} 0 & 0 \\ 0 & 1 \end{pmatrix} = I > 0. \end{aligned}$$

Thus, assumption (H) holds for this example. Next, we compute

$$(4.3) \quad \begin{aligned} A - BR^{-1}B^T G &= \begin{pmatrix} 0 & 1 \\ 0 & 0 \end{pmatrix} - \begin{pmatrix} 1 & 0 \\ 0 & 0 \end{pmatrix} \begin{pmatrix} 1 & 0 \\ 0 & -1 \end{pmatrix} \begin{pmatrix} 1 & 0 \\ 0 & 0 \end{pmatrix} \begin{pmatrix} -1 & 0 \\ 0 & 2 \end{pmatrix} \\ &= \begin{pmatrix} 0 & 1 \\ 0 & 0i \end{pmatrix} - \begin{pmatrix} -1 & 0 \\ 0 & 0 \end{pmatrix} = \begin{pmatrix} 1 & 1 \\ 0 & 0 \end{pmatrix}, \\ BR^{-1}B^T &= \begin{pmatrix} 1 & 0 \\ 0 & 0 \end{pmatrix} \begin{pmatrix} 1 & 0 \\ 0 & -1 \end{pmatrix} \begin{pmatrix} 1 & 0 \\ 0 & 0 \end{pmatrix} = \begin{pmatrix} 1 & 0 \\ 0 & 0 \end{pmatrix}, \\ Q + GA + A^T G - GBR^{-1}B^T G &= \begin{pmatrix} 1 & 1 \\ 1 & 0 \end{pmatrix} + \begin{pmatrix} -1 & 0 \\ 0 & 2 \end{pmatrix} \begin{pmatrix} 0 & 1 \\ 0 & 0 \end{pmatrix} \\ &+ \begin{pmatrix} 0 & 1 \\ 0 & 0 \end{pmatrix} \begin{pmatrix} -1 & 0 \\ 0 & 2 \end{pmatrix} - \begin{pmatrix} 1 & 0 \\ 0 & 0 \end{pmatrix} = 0. \end{aligned}$$

Thus,

$$(4.4) \qquad \mathcal{A} = \begin{pmatrix} 1 & 1 & 1 & 0 \\ 0 & 0 & 0 & 0 \\ 0 & 0 & 1 & 0 \\ 0 & 0 & -1 & 0 \end{pmatrix}, \ \mathcal{C}_1 = \begin{pmatrix} 0 & 0 \\ 0 & 1 \\ 1 & 0 \\ 0 & 1 \end{pmatrix}.$$

Consequently,

$$(4.5) \qquad e^{\mathcal{A}t} = \begin{pmatrix} e^t & e^t - 1 & te^t & 0 \\ 0 & 1 & 0 & 0 \\ 0 & 0 & e^t & 0 \\ 0 & 0 & 1 - e^t & 1 \end{pmatrix}, \quad t \geq 0.$$

We now look at the condition (3.24). Note that in the present case,

$$(4.6) \qquad (0 \ \ I) \, e^{\mathcal{A}(t-s)} \begin{pmatrix} 0 \\ I \end{pmatrix} = (0 \ \ I) \, e^{\mathcal{A}(t-s)} \mathcal{C}_1 = \begin{pmatrix} e^{t-s} & 0 \\ 1 - e^{t-s} & 1 \end{pmatrix}.$$

Hence, (3.24) holds. Therefore, the associated Problem (LQ) is uniquely solvable at any $s \in [0, T]$. The state equation for this problem takes the form:

$$(4.7) \qquad \begin{cases} dx_1(t) = [x_2(t) + u_1(t)]dt \\ dx_2(t) = u_2(t)dw(t), \\ x_1(s) = \xi_1, \quad x_2(s) = \xi_2, \end{cases}$$

and the cost functional is

$$(4.8)$$
$$J(s, \xi; u(\cdot)) = E\Big\{ \int_s^T \big[x_1(t)^2 + 2x_1(t)x_2(t) + u_1(t)^2 - u_2(t)^2 \big] dt - x_1(T)^2 + 2x_2(T)^2 \Big\}.$$

We see that in the above example, B and D are nonzero, Q, R, and G are all indefinite. It is not very hard to cook up many other examples of similar nature.

To conclude this paper, we would like to point out that the case that we have discussed is still very special. The general case is left widely open.

References

[1] J.-M. Bismut, *Linear quadratic optimal stochastic control with random coefficients*, SIAM J. Control Optim., 14 (1976), 419–444.

[2] J. M. Bismut, *An introductory approach to duality in stochastic control*, SIAM Rev., 20 (1978), 62–78.

[3] S. Chen, X. Li, and X. Y. Zhou, *Stochastic linear quadratic regulators with indefinite control weight costs*, SIAM J. Control Optim., 36 (1998), 1685–1702.

[4] S. Chen and J. Yong, *Stochastic linear quadratic optimal control problems*, Appl. Math. Optim., to appear.

[5] S. Chen and J. Yong, *Stochastic linear quadratic optimal control problems II*, submitted.

[6] S. Chen and X. Y. Zhou, *Stochastic linear quadratic regulators with indefinite control weight costs II*, preprint.

[7] R. E. Kalman, *Contributions to the theory of optimal control*, Bol. Soc. Math. Mexicana, 5 (1960), 102–119.

[8] I. Karatzas, and S. Shreve, *Brownian Motion and Stochastic Calculus*, Springer-Verlag, Berlin, 1988.

[9] N. N. Krasovskii, *Stabilization of systems in which noise is dependent on the value of the control signal*, Eng. Cybern. (USSR), no.3 (1965), 94–102.

[10] J. Ma, and J. Yong, *Forward-Backward Stochastic Differential Equations and Their Applications*, Springer-Verlag, Berlin, 1999.

[11] P. J. McLane, *Optimal stochastic control of linear systems with state- and control- dependent disturbances, IEEE Trans. Automat. Control*, AC-16 (1971), 793–798.

[12] E. Pardoux, and S. Peng, *Adapted solutions of backward stochastic equations, Syst. Contr. Lett.*, 14 (1990), 55–61.

[13] W. M. Wonham, *On a matrix Riccati equation of stochastic control*, SIAM J. Control, 6 (1968), 312–326.

[14] J. Yong, *Linear forward-backward stochastic differential equations, Appl. Math. Optim.*, 39 (1999), 93–119.

[15] J. Yong, and X. Y. Zhou, *Stochastic Controls: Hamiltonian Systems and HJB Equations*, Springer-Verlag, New York, 1999.

DEPARTMENT OF APPLIED MATHEMATICS, ZHEJIANG UNIVERSITY, HANGZHOU 310027, CHINA.

LABORATORY OF MATHEMATICS FOR NONLINEAR SCIENCES, DEPARTMENT OF MATHEMATICS, AND INSTITUTE OF MATHEMATICAL FINANCE, FUDAN UNIVERSITY, SHANGHAI 200433, CHINA.

AMS/IP Studies in Advanced Mathematics
Volume 26, 2002

Convertible bonds with market risk and credit risk

Mark Davis and Fabian R. Lischka

ABSTRACT. The incorporation of credit risk in the valuation of convertible bonds has mostly been rather ad-hoc in the literature. Here a model is introduced that attempts a consistent treatment of equity and credit risk.

It incorporates a Black-Scholes stock price (equity risk), Hull-White short rate (interest rate risk), and a hazard rate, depending on the asset price, which determines the probability of default (credit risk). The model can be calibrated to match the initial term structure of interest rates as well as the 'term structure of credit spreads'.

CONTENTS

1.	Introduction	45
1.1.	Convertible bonds	46
1.2.	Valuation by arbitrage: The classical literature	47
1.3.	Dealing with credit risk:	
	The modern literature	48
2.	Underlyings: Stock price, short rate, hazard rate	50
2.1.	One factor	51
2.2.	Two factors: Stochastic interest rate	51
2.3.	Two factors: Stochastic hazard rate	53
2.4.	Two-and-a-half factors	53
3.	Empirical results	56
	References	58

1. Introduction

A convertible bond is a coupon paying corporate bond that can be converted into company stock at the discretion of the holder. The purpose of this article is to describe a method to value a convertible bond, incorporating both interest rate and credit risk.

A convertible bond is a challenging instrument to value, because it is both an equity and an interest rate derivative. These two components are subject to different credit risk, because a company can always issue more of its stock, but not necessarily come up with sufficient cash to meet bond obligations. Furthermore, in reality, even the most basic convertible bonds often incorporate various additional features, such as call and put provisions, strike reset features, mandatory or restricted conversion, etc.

Three sources of randomness are at work: the stock price, the interest rate, and the credit spread. The probability of default over the next small period is given by the hazard rate. One could model all factors at the same time, but practitioners tend to eschew models with more than two factors.

There are various ways of reducing the problem to two factors: either, take the hazard rate to be a prespecified deterministic function of time, and model stock price and interest rate stochastically. Alternatively, assume the interest rate to be deterministic and the hazard rate to be stochastic.[1]

A third way will be explored here: A two dimensional trinomial lattice will be used to model the (rebased) stockprice on one dimension and the interest rate on the other. The hazard rate is assumed to be a deterministic function of the stock price: if the stock price falls below its (risk neutral) expectation, the hazard rate rises, and vice versa.

1.1. Convertible bonds. Convertible bonds are typically listed securities issued by companies and traded on secondary markets. The ratio of convertible to total debt was above 10% on average in the United States during this century [**9**, p. 233], and they accounted for above 5on the London Stock Exchange from 1973 to 1995 [**11**]. Companies issue them mainly because they enable them to lower their costs of debt funding ('debt sweetener') by implicitly selling an option on their stock (which will only be exercised if the company is doing well). This helps to resolve the problem of asymmetric information on the riskiness of the underlying assets, and reduces agency costs.[2] The other commonly cited reason for issuing convertibles is as an indirect and delayed issuing of equity ('delayed equity'), reducing dilution and circumventing regulatory hindrances.[3]

Investors on the other hand hold coonvertibles for their upside potential with limited downside. Sophisticated investors tend to consider convertibles a good deal, particularly considering the time value of the embedded option.[4]

As long as the bond holder does not convert, he regularly receives a coupon and is finally repaid his principal. If he chooses to convert during the lifetime of the bond, however, the bond is redeemed and the issuer receives some ordinary shares of the company at a previously agreed exchange price of EP per share.[5] Since the notional K of the bond is assumed to be converted upon redemption, the value received on conversion—parity—is $K\frac{S_t}{EP}$. This effectively constitutes a call option

---

[1]These two approaches seem very symmetrical, particularly since in the first case, a risk-free zero coupon bond is used as numeraire, and in the second a 'defaultable' dollar, hence the implementations are very similar. However, the treatment of the recovery value differs.

Both approaches have been explored and implemented by Davis [**3**].

[2]Equity holders have an incentive to increase the riskiness of the assets, which increases the value of equity and decreases the value of debt (holding firm value constant). On the other hand, the call component of a convertible increases in value, so the total value of the convertible can be made insensitive to changes in risk.

[3]Nyborg [**11**] explores in greater depth the motives and strategic issues involved in the issue of convertible debt.

[4]Kang and Lee [**9**] provide an empirical investigation of convertible debt offerings and report significant initial underpricing.

[5]The exchange price is typically set a bit higher than the current stock price to avoid immediate conversion.

FIGURE 1. Payoff of a convertible bond (with $\kappa = 1$).

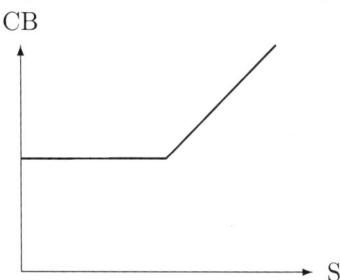

on the stock, since the value of the convertible bond CB at maturity is roughly

$$(1) \qquad CB = \max(B, \kappa S) = B + (\kappa S - B)^+,$$

where $\kappa = K/EP$ is the conversion ratio, see figure 1.

The bond pays coupons, the stock pays dividends. Hence there are intermediate cash flows, some of which are stochastic, though that is rarely modelled.

Features of convertible bonds. Very often, convertibles are callable: The issuer has the option to buy the bond back at a predetermined strike price (Often this strike price changes during the life of the bond). However, the holder typically has the right to convert the bond rather than deliver it once the call announcement has been made, hence the call provision is often used to force early conversion of the bond.

Early conversion of a convertible bond is never optimal under certain conditions[6], hence this call provision reduces the value of the convertible: It limits the investor's return if interest rates fall or the stock price rises. Often, convertible bonds are call-protected for some years and become callable only after that.

On the other hand, a put provision allows the holder to return the convertible to the issuer in exchange for cash at certain points in time for a predetermined price, and hence offer additional downside protection in case of rising interest rates.

With even more complex bonds, e.g. Japanese resettable convertibles, the holder's conversion options change according to the stock price and its history.[7]

1.2. Valuation by arbitrage: The classical literature. Theoretical pricing models for convertible bonds first appeared in the 1960's,[8] though they were severely limited because they typically considered only one particular point in the future, found the maximum of conversion and bond value, and discounted at some rate.

Shortly after the breakthrough in option pricing by Black and Scholes in 1973 (and Merton's 'rational' option pricing theory same year), the new, theoretically sound method was applied to the case of convertible bonds.

---

[6]See Ingersoll [**7**, Theorem I] for the exact conditions (essentially perfect market, no dividends, and constant conversion ratio).

[7]See RISK magazine [**12**] for these and more complications. These complications make the convertible path dependent and hence indicate Monte-Carlo simulation as the appropriate modelling tool. We ignore these possibilities, and since convertibles are American style instruments, modelling on a tree seems appropriate.

[8]See [**7**] for references.

Ingersoll's paper [7] from 1976 develops arbitrage arguments to derive several results concerning the optimal conversion strategy (for the holder) and call strategy (for the issuer) as well as closed form solutions for the value of convertible bonds in a variety of special cases. He assumes that the value of the firm as a whole is composed of equity and convertible bonds only and follows a geometric Brownian motion.

Brennan and Schwartz [1] publish similar results in 1977 under the same assumptions. They develop the PDE and boundary conditions for the value of convertible bonds under fairly general conditions and describe a finite difference method to solve it.

Credit risk in early approaches. In case the firm value falls under the debt notional, the firm defaults and the bond holders aquire the assets of the firm.

In this sense, credit risk is taken into account in these early papers. This approach considers the capital structure of the firm and views the convertible bond not as a derivative on the stock price, but a compound option on the physical assets underlying the financial securities, similar in spirit to Merton's 1977 model of debt as a portfolio of notional and a short put option on the total firm value. However, this theoretically appealing methodology suffers from problems in practice, as pointed out in 1995 by Jarrow and Turnbull [8]: The underlying physical assets are often not tradable, and their value and volatility not necessarily known. Second, all corporate liabilities senior to the convertible bond at hand must be valued at the same time. These complications make this approach unsuitable for pricing corporate liabilities[9] in practice.

1.3. Dealing with credit risk:

The modern literature. Newer approaches therefore consider the convertible bond as a security contingent on the stock price and interest rate. The stock price is assumed to follow a geometric Brownian motion. A tree to value a convertible bond, even with call and put provisions, coupons, dividends, and other features, can then be constructed in the usual way.

The essential difficulty is the choice of the discount rate. Suppose the convertible is certain to remain a bond. The cash payoff is then subject to credit risk, and the appropriate interest rate incorporates a credit spread corresponding to the credit rating of the issuer.

Suppose on the other hand the convertible is certain to be converted. The firm can always issue a share, which can then be sold and the proceeds invested risk free. The appropriate discounting rate is then the risk free rate.

Adjusting the credit spread. One way this issue has been dealt with is detailed in a research note [4] from Goldman Sachs (1994), and outlined in Hull [6, section 20.5]. They consider the probability of conversion at every node, and choose the discount rate to be an accordingly weighted arithmetic average between risk free rate and the risky rate (which is obtained by adding the issuer's credit spread).

The probability of conversion at the final layer (t=T) of the tree is either 1 or 0, depending on whether the convertible is converted or not. At previous nodes, the conversion probability is calculated as the average of the conversion probabilities of the successor nodes. If the convertible is converted (or put) at a node, the conversion probability is reset to 1 (or 0). Goldman Sachs describe a tree using this method in greater detail, including call and put provisions.

---

[9]It is more promising for areas where these problems are less prevalent, e.g. mortgages.

FIGURE 2. Decomposing the payoff

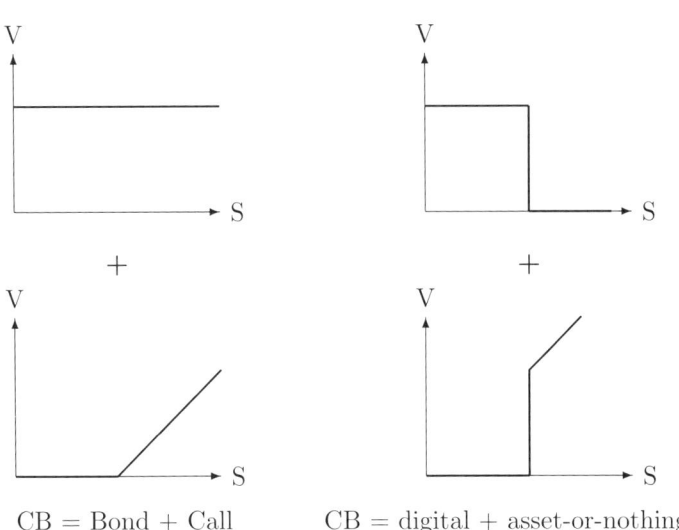

CB = Bond + Call   CB = digital + asset-or-nothing

Separating cash and equity. A different approach was introduced by Tsiveriotis and Fernandes [**13**] in 1998. Rather then considering the convertible bond value as a portfolio of normal bond and call option (see figure 2, left column), they decompose it in a cash account and stock account (right column). Both are maintained and discounted seperately.[10] This is done by introducing an artificial security, namely the cash only part of the convertible bond, that pays all the cash payments, but no equity that an optimally behaving holder of a convertible would receive. They derive the joint PDE and boundary conditions for the artificial security and the normal convertible and describe the numerical solution using the explicite finite difference method.

All models mentioned so far do not consider interest rate risk, which for convertible bonds is evidently more important than for pure equity options. Brennan and Schwartz (1980, [**2**]) and Longstaff and Schwartz (1993, [**10**]) introduce two factor models that include interest rate risk, but they are not necessarily consistent with the initial term structure of interest rates.

Ho and Pfeffer [**5**] analysed a two factor model in 1996 that adapts to the initial term structure. When the stock movement is ignored, their two dimensional binomial lattice produces the same interest rates and arbitrage free bond prices as in the Ho and Lee model (1986). They analyse a sample of bonds and place special emphasis on hedging strategies and calculating the greeks. However, they consider bond value ('investment value') and call value ('Latent Warrant') separately, and capture the credit risk by a constant credit spread that is added to the Treasury rate at each point to discount the bond's cash flow. Hence, their approach might be very valuable for investors, but is not very satisfying theoretically.

---

[10]This seems similar to the previously mentioned method, in which the probability of conversion is calculated and an accordingly adjusted discount rate used. However, that fails to account correctly for intermediate (risky) cash flows, such as coupons or contingent flows due to call and put provisions.

The present approach: hazard rate depending on stock price. The approach described here incorporates a Black-Scholes asset price, Hull-White short rate, and a hazard rate that determines the probability of default. The model can be calibrated to match the initial term structure of interest rates as well as the 'term structure of credit spreads', i.e. the implied (risk neutral) survival probability of a company.[11]

Hence, the model is consistent with observed market data, poses no unsurmountable problems concerning parameter estimation, is theoretically appealing, can be calculated straightforwardly on a two dimensional lattice, and there seem to be no theoretical reasons precluding the incorporation of more realistic features of convertible bonds.

2. Underlyings: Stock price, short rate, hazard rate

Default behaviour. The default is modelled by a point process $N(t)$, which starts out at zero but jumps to one when default occurs. When default occurs, the stock price jumps to zero[12], while the price of the coupon bearing convertible bond jumps to a recovery value ℓK, a predetermined fraction of the notional.

The probability of default over a small period is proportional (to first order) to the time varying intensity $h(t)$, the hazard rate. Then the expectation of $N(t)$ under $\mathbb{Q}$ is[13] $E[N(t)] = \int_{s=0}^{t} h(s)\,ds$, so

$$dM(t) = dN(t) - \big(1 - N(t)\big)h(t)\,dt$$

defines a martingale.

The hazard rate is modelled as deterministic or stochastic, depending on the model, and can be calibrated to reproduce the implied survival probabilities[14].

Stock price in general. The stock price follows

$$\begin{aligned}
dS(t) &= \big(r(t) - y(t)\big)S(t)\,dt + \sigma_1 S(t)\,dW_1(t) \\
&\quad - S(t_-)\big(dN(t) - h(t)\,dt\big) \\
&= \big[r(t) + h(t) - y(t)\big]S(t)\,dt + \sigma_1 S(t)\,dW_1(t) - S(t_-)\,dN(t).
\end{aligned}$$

It is a martingale under $\mathbb{Q}$ when rebased by the money market account. When default occurs, the stock price immediately before default $S(t_-)$ is subtracted, i.e. the stock price jumps to zero and stays there. Prior to default, the stock price is given by

$$(2) \qquad S(t) = S_0 \exp\left[\int_0^t \Big(r(s) + h(s) - y(s)\Big)\,ds - \frac{1}{2}\sigma_1^2 t + \sigma_1 W_1(t)\right].$$

---

[11]It can—under certain assumptions—be recovered from comparing prices of company bonds with Treasury bonds, see [**8**].

[12]Note that the jump size is previsible, so it is possible to hedge against default. This enables the arbitrage argument.

[13]Questions of existence and uniqueness of the equivalent martingale measure are deferred for the time being. Note that the survival probability is not a tradable asset and need not grow at the riskless rate.

One could introduce a defaultable zero coupon bond (issued by the issuer of the shares) that would need to be a martingale when appropriately rebased, and develop hedging strategies along that line. However, a defaultable bond turns out not to be log-normal, hence a different approach is used here.

[14]They can be estimated from market data (namely the credit spread of the issuing company over Treasury), once an assumption concerning the recovery value has been made, see Jarrow and Turnbull (1995,[**8**]).

The risk neutral stock price return is increased by $h(t)$ to compensate for the possibility of default and reduced by $y(t)$ to compensate for dividend payments.

2.1. One factor. Short rate and hazard rate are first assumed to be deterministic functions of time. Then one can write (2) as

$$S(t) = F_A(t) \cdot \mathcal{E}[\sigma_1 W_1](t),$$

where $F_A(t) = S_0 \exp\left[\int_0^t \left(r(s) + h(s) - y(s)\right) ds\right]$ is a deterministic function of time which can be interpreted as the forward price, and

$$\mathcal{E}[\sigma_1 W_1](t) = \frac{e^{\sigma_1 W_1(t)}}{E\left[e^{\sigma_1 W_1(t)}\right]} = e^{\sigma_1 W_1(t) - \frac{1}{2}\sigma_1^2 t}$$

is the exponential martingale of the martingale $\sigma_1 W_1$.

Hence, for calculations, one just has to calculate the forward $F_A(t)$, model $\mathcal{E}[\sigma W_1]$ on a one-dimensional tree, and can then immediately calculate the stock price at every node and hence derivatives prices.

2.2. Two factors: Stochastic interest rate.

Short rate dynamics. The hazard rate is still assumed to be deterministic, while the short rate dynamics are now modelled by a mean reverting [15] process with time varying mean (extended Vasicek):

$$(3) \qquad dr(t) = \left[\theta(t) - \lambda\, r(t)\right] dt + \sigma_2\, dW_2(t),$$

where $W_2(t)$ is a Brownian motion with instantaneous correlation ρ to the first Brownian motion, all under the risk neutral measure $\mathbb{Q}$.

The mean reversion parameter λ and the volatility σ_2 are assumed to be constant, while the $\theta(t)$ is assumed to be a (locally bounded) deterministic function of time, used to calibrate to the observed term structure of interest rates.

The unique solution to (3) is given by the Gaussian process

$$(4) \qquad r(t) = e^{-\lambda t} r_0 + e^{-\lambda t}\int_0^t e^{\lambda s}\theta(s)\, ds + e^{-\lambda t}\sigma_2 \int_0^t e^{\lambda s}\, dW_2(s)$$

with mean and covariance function

$$\alpha(t) := E_{\mathbb{Q}}[r(t)] = e^{-\lambda t} r_0 + e^{-\lambda t}\int_0^t e^{\lambda s}\theta(s)\, ds$$

$$\mathrm{Cov}[r(s), r(t)] = \frac{\sigma_2^2}{2\lambda}\left(e^{-\lambda|s-t|} - e^{-\lambda(s+t)}\right).$$

Observe that the first two terms in (4) (the expectation of $r(t)$) are just a deterministic function of time, while the remaining term is a Gaussian martingale. Hence write

$$(5) \qquad\qquad\qquad r(t) = \alpha(t) + e^{-\lambda t} X_2(t),$$

$$(6) \qquad\qquad \text{where} \qquad X_2(t) := \sigma_2 \int_0^t e^{\lambda s}\, dW_2(s).$$

---

[15] The term 'mean reverting' is slightly misleading, since the specification (3) below is under the riskneutral, not the real world measure. How the process behaves under the change of drift concomitant with the change of measure is unspecified. However, it is sure to say that the short rate has finite unconditional variance.

One does not need to bother with $\theta(t)$ any further, but can immediately choose $\alpha(t)$ to fit the current zero coupon bond prices.

Bond prices. Under the risk neutral measure $\mathbb{Q}$, the price at time t of the bond maturing at time T is given by $P^T(t) = E_{\mathbb{Q}}\left[e^{-\int_t^T r(s)\,ds}\Big|\mathcal{F}_t\right]$, and Feynman-Kac yields the usual PDE with solution

$$P^T(t) = e^{A^T(t) - B^T(t)r(t)},$$

$A^T(t)$ is not further needed. Solving for $B^T(t)$ gives

$$(7) \qquad\qquad B^T(t) = \frac{1}{\lambda}\left(1 - e^{-\lambda(T-t)}\right)$$

Finally, plugging the obtained values in Itô's lemma yields the process for the T-bond price under $\mathbb{Q}$:

$$\frac{dP^T(t)}{P^T(t)} = r(t)\,dt - \sigma_2\,B^T(t)\,dW_2(t),$$

The stock price with stochastic interest rates. Again, the stock price prior to default is

$$S(t) = S_0 \exp\left[\int_0^t \left(r(s) + h(s) - y(s)\right)ds - \frac{1}{2}\sigma_1^2 t + \sigma_1 W_1(t)\right].$$

However, now the integral over $r(t)$ is stochastic and path dependent. This complicates numerical calculations because it precludes filling out the tree from the last ($t = T$) layer. Hence, consider the T-forward price: rebase the asset with respect to a zero coupon bond with maturity time T and consider $Y_B(t) = S(t)/P^T(t)$. Itô's Lemma yields

$$(8) \qquad \begin{aligned} \frac{dY_B(t)}{Y_B(t)} &= \left(\left(\sigma_2 B^T(t)\right)^2 + \rho\sigma_1\sigma_2 B^T(t) - y(t)\right)dt \\ &\qquad + \sigma_1\,dW_1(t) + \sigma_2 B^T(t)\,dW_2(t) - dN(t) \\ &=: \left(m(t) - y(t)\right)dt + \tilde{\sigma}(t)\,d\tilde{W}(t) - dN(t). \end{aligned}$$

The stochastic short rate $r(t)$ canceled out, and the remaining drift term

$$m(t) = \left(\sigma_2 B^T(t)\right)^2 + \rho\sigma_1\sigma_2 B^T(t)$$

is a deterministic function of time.

The two random parts in line (8) stemming from asset and interest risk respectively can be aggregated into

$$dX_1(t) = \sigma_1\,dW_1(t) + \sigma_2 B^T(t)\,dW_2(t)$$

$$X_1(t) = \int_0^t \tilde{\sigma}(s)d\tilde{W}(s),$$

where the aggregated instantaneous volatility $\tilde{\sigma}(t)$ is

$$\tilde{\sigma}(t) = \sqrt{\sigma_1^2 + 2\rho\sigma_1\sigma_2 B^T(t) + \left(\sigma_2 B^T(t)\right)^2},$$

and the new Brownian motion $\tilde{W}(t)$ is defined by

$$\tilde{\sigma}(t)d\tilde{W}(t) = \sigma_1\,dW_1(t) + \sigma_2 B^T(t)\,dW_2(t).$$

With these definitions, $\frac{dY_B(t)}{Y_B(t)} = \big(m(t) - y(t)\big)\, dt + dX_1(t)$, so Y_B is log-normal and

$$Y_B(t) = F_B(t) \cdot \mathcal{E}[X_1](t)$$

with $F_B(t) = Y_1(0) \exp\left[\int_0^t \big(m(s) - y(s)\big)\, ds\right]$. Hence $Y_B(t)$ is again the product of a deterministic forward and an exponential martingale $\mathcal{E}[X_1]$.

The short rate $r(t) = \alpha(t) + e^{-\lambda t} X_2(t)$, on the other hand, is a deterministic function of time and the other martingale X_2.

To model the twodimensional Gaussian martingale on a two dimensional tree, one needs the (time varying) covariance structure.

Write

$$\mathrm{Var}\left[\mathbf{X}(t) = \begin{pmatrix} X_1(t) \\ X_2(t) \end{pmatrix}\right] = \int_0^t Q(s)ds = t \cdot \bar{Q}(t),$$

where $Q(t)$ contains 'forward' or 'instantaneous' covariances while $\bar{Q}(t)$ contains 'term' or 'average' covariances. Then

$$Q_{11}(t) = \sigma_1^2 + 2\rho\sigma_1\sigma_2 B^T(t) + \big(\sigma_2 B^T(t)\big)^2$$
$$Q_{22}(t) = \sigma_2^2 e^{2\lambda t}$$
$$Q_{12}(t) = \rho\sigma_1\sigma_2 e^{\lambda t} + \sigma_2^2 B^T(t)e^{\lambda t}.$$

2.3. Two factors: Stochastic hazard rate. Now, take the short rate to be deterministic, and model the hazard rate as a mean-reverting Ornstein-Uhlenbeck process.

The situation then is symmetric to the model before, the same tree can be used. However, the short rate and hazard rate are exchanged, which results in different treatment of the recovery value in case of default.

The stock price. Again, the stock price prior to default is given by (2). However, now the integral over $h(t)$ is stochastic and pathdependent, hence rebase[16] by the survival probability to maturity $\eta^T(t)$.

2.4. Two-and-a-half factors. Model the short rate as before as extended Vasicek, equation (3), and the hazard rate as[17]

$$\begin{aligned} h(t) &= \gamma(t) - \sigma_3 W_1(t), \text{ so} \\ dh(t) &= \gamma'(t) - \sigma_3\, dW_1(t), \end{aligned}$$

(9)

where W_1 is the Brownian motion that is driving the asset price. The summand $\gamma(t)$ is used to calibrate the hazard rate so that is reproduces the implied survival probabilities[18].

---

[16]This corresponds to the idea of an exchange rate between dollars and 'defaultable dollars' as proposed in 1995 by Jarrow and Turnbull [8].

[17]With a reasonable choice of σ_3 one can prevent the hazard rate from turning negative while the stock return stays in, say, five standard deviations from expectation.

[18]Jarrow and Turnbull [8] show how to estimate the survival probabilities under the risk neutral measure from market data (namely the credit spread of the issuing company over Treasury), once an assumption concerning the recovery value has been made.

Survival probability. The survival probability $\eta^T(t)$ (i.e. the probability that $N(T) = 0$ given $N(t) = 0$) is[19]

$$\eta^T(t) = E\left[e^{-\int_t^T h(s)\,ds}\Big|\mathcal{F}_t\right],$$

so, for the choice of hazard rate above,

$$\eta^T(t) = \exp\left[-\int_t^T \gamma(s)\,ds + \sigma_3 W_1(t)(T-t) + \frac{\sigma_3^2}{6}(T-t)^3\right],$$

and applying Itô's lemma, one obtains the SDE for the survival probability:

$$(10) \qquad \frac{d\eta^T(t)}{\eta^T(t)} = h(t)\,dt + \sigma_3 \cdot (T-t)\,dW_1(t).$$

The survival probability grows with rate $h(t)$, analogue to the bond.

Again, the stock price prior to default is given by (2). However, now the integral over *both* $r(t)$ and $h(t)$ is stochastic and pathdependent, hence rebase[20] by both bond and survival probability, $Y_D(t) = S(t)/(P^T(t) \cdot \eta^T(t))$.

$$\frac{dY_D(t)}{Y_D(t)} = \left[\left(\sigma_2 B^T(t)\right)^2 + \sigma_3^2(T-t)^2 + \rho\sigma_1\sigma_2 B^T(t)\right.$$

$$\left. - \rho\sigma_2\sigma_3 B^T(t)(T-t) - \sigma_1\sigma_3(T-t) - y(t)\right]dt$$

$$(11) \qquad + [\sigma_1 - \sigma_3(T-t)]\,dW_1(t) + \sigma_2 B^T(t)\,dW_2(t) - dN(t).$$

Both the stochastic short rate $r(t)$ and hazard rate $h(t)$ cancelled out, and the drift term

$$m(t) = \left(\sigma_2 B^T(t)\right)^2 + \sigma_3^2(T-t)^2 + \rho\sigma_1\sigma_2 B^T(t)$$

$$- \rho\sigma_2\sigma_3 B^T(t)(T-t) - \sigma_1\sigma_3(T-t)$$

is a deterministic function of time that can be calculated right away.

The random parts in line (11) stemming from asset, default and interest risk respectively can be aggregated into

$$dX_1(t) = [\sigma_1 - \sigma_3(T-t)]\,dW_1(t) + \sigma_2 B^T(t)\,dW_2(t).$$

With this definition, $\frac{dY_D(t)}{Y_D(t)} = [m(t) - y(t)]\,dt + dX_1(t) - dN(t)$, so Y_D is log-normal and

$$(12) \qquad Y_D(t) = F_D(t) \cdot \mathcal{E}[X_1](t)$$

with $F_D(t) = Y_D(0)\exp\left[\int_0^t (m(s) - y(s))\,ds\right]$. Hence $Y_D(t)$ is again the product of a deterministic forward and an exponential martingale $\mathcal{E}[X_1]$.

---

[19]Analogue to zero coupon prices as a function of the short rate.

[20]It might seem more natural to use a defaultable zero coupon bond as a numeraire. Assuming that its value jumps to a predetermined fraction of a non defaultable zero coupon bond upon default, however, it turns out that the defaultable zero is not log-normal.

Volatilities and Covariances. For modelling, the covariance structure is needed. Write

(13)
$$\mathrm{Var}\begin{pmatrix} X_1(t) \\ X_2(t) \\ W_1(t) \end{pmatrix} = \int_0^t Q(s)ds = t \cdot \bar{Q}(t).$$

With

$$dX_1(t) = [\sigma_1 - \sigma_3(T - t)]\, dW_1(t) + \sigma_2 B^T(t)\, dW_2(t)$$
$$dX_2(t) = \sigma_2\, e^{\lambda t}\, dW_2(t),$$

we have

$$Q_{11}(t) = \left(\sigma_1 - (T-t)\sigma_3\right)^2 + 2\rho\left(\sigma_1 - (T-t)\sigma_3\right)\sigma_2 B^T(t) + \left(\sigma_2 B^T(t)\right)^2$$
$$Q_{22}(t) = \sigma_2^2 e^{2\lambda t}$$
$$Q_{33}(t) = 1$$
$$Q_{12}(t) = \rho\left(\sigma_1 - (T-t)\sigma_3\right)\sigma_2 e^{\lambda t} + \sigma_2^2 B^T(t)e^{\lambda t}$$
$$Q_{13}(t) = \sigma_1 - (T-t)\sigma_3 + \rho\sigma_2 B^T(t)$$
$$Q_{23}(t) = \rho\sigma_2 e^{\lambda t}.$$

and, with the abbreviations $C(T,t) = \frac{e^{-\lambda(T-t)} - e^{-\lambda T}}{\lambda t}$ and $c(T) = C(T,T) = \frac{1 - e^{-\lambda T}}{\lambda T}$:

$$\bar{Q}_{11}(t) = \sigma_1^2 + \frac{2\rho\sigma_1\sigma_2 + \sigma_2^2}{\lambda}\left(1 - C(T,t)\right)$$
$$\qquad - \frac{2\rho\sigma_2\sigma_3}{\lambda^2}\left(e^{-\lambda(T-t)} + \tfrac{\lambda}{2}(2T - t) - (1 + \lambda T)C(T,t)\right)$$
$$\qquad - \sigma_1\sigma_3(2T - t) + \frac{\sigma_3^2}{3}(3T^2 - 3tT + t^2)$$
$$\bar{Q}_{22}(t) = \sigma_2^2 \cdot c(-2t)$$
$$\bar{Q}_{33}(t) = 1$$
$$\bar{Q}_{12}(t) = \frac{\sigma_2^2}{\lambda}\left(c(-t) - C(T, 2t)\right) + \rho\sigma_1\sigma_2 \cdot c(-t)$$
$$\qquad - \frac{\rho\sigma_2\sigma_3}{\lambda}\left((1 + \lambda T)c(-t) - e^{\lambda t}\right)$$
$$\bar{Q}_{13}(t) = \sigma_1 + \frac{\rho\sigma_2}{\lambda}\left(1 - C(T,t)\right) - \frac{\sigma_3(2T - t)}{2}$$
$$\bar{Q}_{23}(t) = \rho\sigma_2 \cdot c(-t)$$

Recovering W_1. To keep the computation tractable, only $X_1(t)$ and $X_1(t)$ are modelled directly on the tree. Hence, $W_1(t)$ is inaccessible.[21] Therefore, a linear estimator is used to obtain its conditional expectation given $\mathbf{X}(t)$:

(14)
$$\hat{W}_1(t) = E[W_1(t)|\mathbf{X}(t)] = \boldsymbol{\beta}'(t)\mathbf{X}(t)$$

with

$$\boldsymbol{\beta}'(t) = \begin{pmatrix} \bar{Q}_{13}(t) \\ \bar{Q}_{23}(t) \end{pmatrix}' \cdot \left[\begin{pmatrix} \bar{Q}_{11}(t) & \bar{Q}_{12}(t) \\ \bar{Q}_{12}(t) & \bar{Q}_{22}(t) \end{pmatrix}\right]^{-1}$$

---

[21] Even if all nodes were stored while going forward (which is unneccessary for this approach), the tree including W_1 would not recombine.

On the other hand, from (2), one obtains

$$\sigma W_1(t) = \ln \frac{S(t)}{F_A(t)} + \tfrac{1}{2}\sigma_1^2 t$$

$$= \ln \frac{S(t)}{S_0} - \int_0^t \big(r(s) + h(s) - y(s)\big)\, ds + \tfrac{1}{2}\sigma_1^2 t$$

$$= \ln \frac{P^T(t)\eta^T(t)}{P^T(0)\eta^T(0)} - \ln E[e^{X_1(t)}]$$

$$+ \int_0^t \big(m(s) - r(s) - h(s)\big)\, ds + \tfrac{1}{2}\sigma_1^2 t + X_1(t)$$

This seems to give $W_1(t)$ as a function of the factor $X_1(t)$, given that one knows the zero coupon value and survival probability at a node. However, the integral over $r(t)$ and $h(t)$ is pathdependent.

Summary of the two-and-a-half factor model. So far, we have expressed the rebased asset price $Y_D(t)$, interest rate and hazard rate as some deterministic (time dependent) functions of the Gaussian martingale $\mathbf{X}(t)$ and $W_1(t)$, the distribution of which is known.

$$Y_D(t) = f_1(X_1(t), t) \qquad\qquad \text{Equation (12)}$$
$$r(t) = f_2(X_2(t), t) \qquad\qquad \text{Equation (5)}$$
$$h(t) = f_3(W_1(t), t) \qquad\qquad \text{Equation (9)}$$

$$\left[\begin{pmatrix} \mathbf{X}(t) - \mathbf{X}(s) \\ W_1(t) - W_1(s) \end{pmatrix} \,\middle|\, \mathcal{F}_s; N(s) = 0 \right] \sim N\left[\mathbf{0}, \int_s^t Q(u)\, du \right]$$

3. Empirical results

If the price of the convertible as a function of current stock price is plotted for different hazard rate volatilities, there is virtually no difference for very small and very large stock prices, due to the calibration to the term structure of interest rates and the term structure of credit spreads respectively.

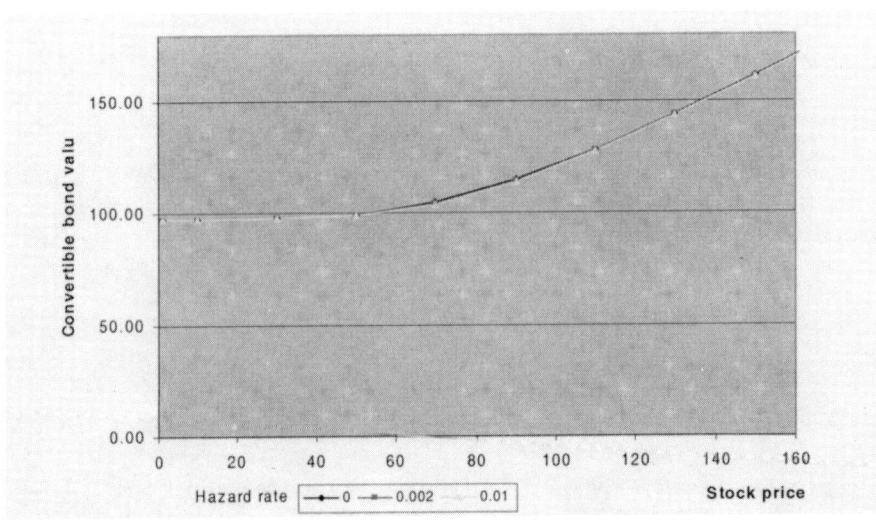

However, for stock prices close to par, the higher the hazard rate volatility, the lower the convertible value. Also, the value of the opportunity to convert prior to maturity ('American' convertible, as opposed to a theoretical 'European') increases as the hazard rate volatility increases.

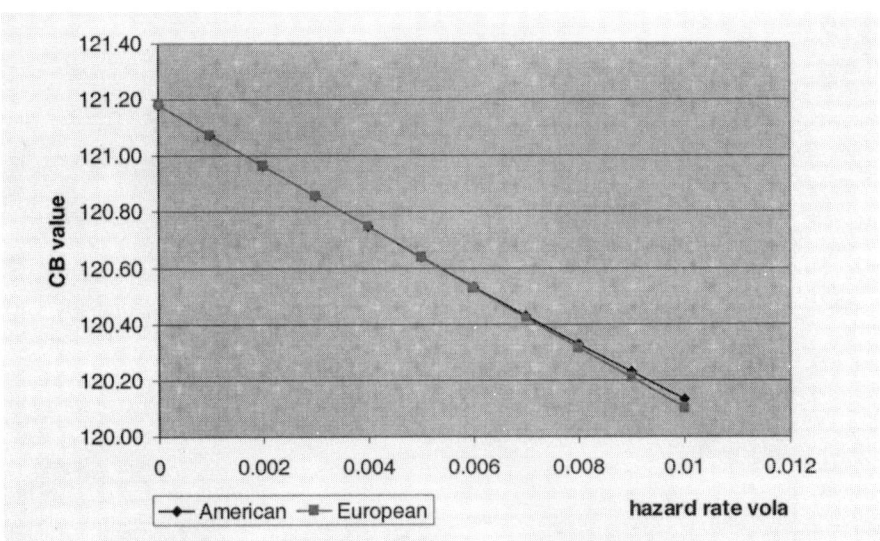

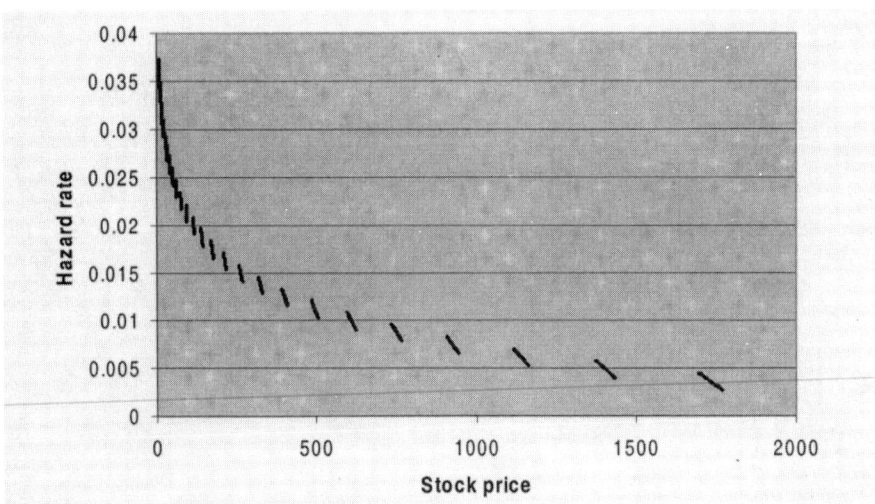

The hazard rate as estimated using $\hat{W}_1(t)$ from equation (14) behaves as intuitively expected, it approaches zero[22] as the stock price increases, and increases rapidly as the stock price falls to zero. Plotting the hazard rate against the logarithm of the stock price reveals the ultimately linear relationship. The hazard rate depends on W_1 not X_1 and X_2, W_1 is a linear estimator using X_1 and X_2, and the deviations from the line come from the information X_2 provides.

---

[22]Only if σ_3 is carefully chosen, otherwise it might turn negative.

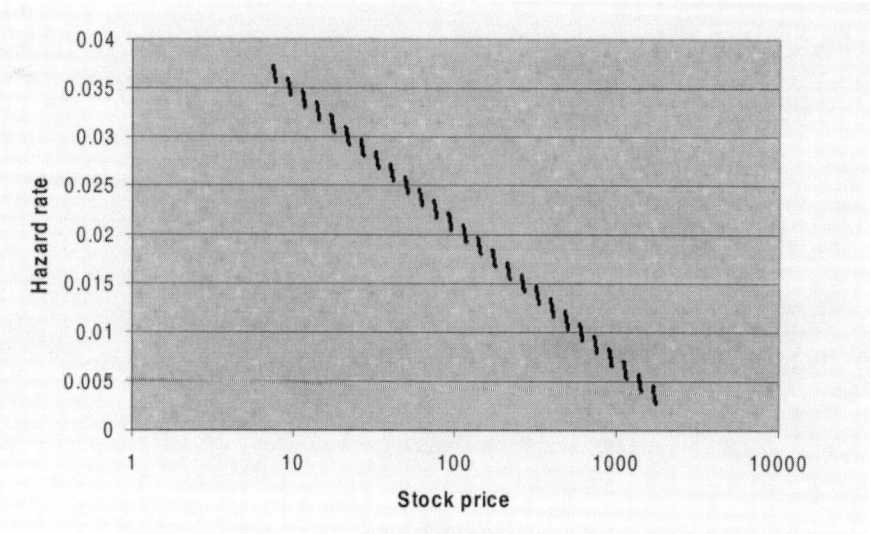

References

[1] M. J. Brennan and E. S. Schwartz. Convertible bonds: Valuation and optimal strategies for call and conversion. *Journal of Finance*, 32(5):1699–1715, December 1977.

[2] M. J. Brennan and E. S. Schwartz. Analyzing convertible bonds. *Journal of Financial and Quantitative Analysis*, 15(4):907–929, November 1980.

[3] Mark Davis. Valuation of convertible bonds. Technical report, Tokyo-Mitsubishi International, London, February 1999.

[4] Goldman Sachs. Valuing convertible bonds as derivatives. Quantitative Strategies Research Notes, Goldman Sachs, New York, November 1994.

[5] Thomas S. Y. Ho and David M. Pfeffer. Convertible bonds: model, value attribution, and analytics. *Financial Analysts Journal*, pages 35–44, September 1996.

[6] John C. Hull. *Options, Futures, and Other Derivatives*. Third edition, 1997.

[7] Jonathan E. Ingersoll, Jr. A contingent-claims valuation of convertible securities. *Journal of Financial Economics*, 4(3):289–321, 1977.

[8] Robert A. Jarrow and Stuart M. Turnbull. Pricing derivatives on financial securities subject to credit risk. *Journal of Finance*, 50(1):53–85, March 1995.

[9] Jun-Koo Kang and Yul W. Lee. The pricing of convertible debt offerings. *Journal of Financial Economics*, 41:231–248, 1996.

[10] F. Longstaff and E. Schwartz. Valuing risky debt: A new approach. Working paper, University of California at Los Angeles, 1993.

[11] K. G. Nyborg. The use and pricing of convertible bonds. *Applied Mathematical Finance*, 3:167–190, 1996.

[12] The challenge of equity derivatives. *RISK, Technology supplement*, pages S29–S31, August 1999.

[13] Kostas Tsiveriotas and Chris Fernandes. Valuing convertible bonds with credit risk. *Journal of Fixed Income*, 8(2):95–102, September 1998.

TOKYO-MITSUBISHI INTERNATIONAL PLC, 6 BROADGATE, LONDON EC2M 2AA

AMS/IP Studies in Advanced Mathematics
Volume 26, 2002

QUASI-MONTE CARLO METHODS AND THEIR RANDOMIZATIONS

FRED J. HICKERNELL and HEE SUN HONG

ABSTRACT. Quasi-Monte Carlo methods are a way of improving the efficiency of Monte Carlo methods. Randomized quasi-Monte Carlo methods combine the advantages of Monte Carlo and quasi-Monte Carlo methods. This article introduces some of the important theoretical and practical ideas underlying randomized quasi-Monte Carlo methods, and explains how these methods can be applied to finance problems.

1. Introduction

Monte Carlo methods are one of the main computational tools for solving problems in mathematical finance. Unfortunately, the convergence of Monte Carlo methods is relatively slow. Quasi-Monte Carlo methods can speed the convergence of Monte Carlo methods and several investigators have found them to be useful in practice for finance computations [**PT95, NT96, PT96, CMO97, OT97, ABG98, LL98, Tez98**]. However, deterministic quasi-Monte Carlo methods have certain disadvantages as well. For example, they do not facilitate simple error estimates. Also, their deterministic nature means that they are biased in a statistical sense, so they may give poor accuracy for some particular problems. In this article we explain how randomized quasi-Monte Carlo methods combine the advantages of both Monte Carlo and quasi-Monte Carlo methods.

1.1. Why Use Monte Carlo Methods for Finance Applications? Let S_i denote the price of a security at the end of the ith trading period. Suppose that one would like to price a derivative based on this stock, such as a put or call option, that has a strike price K, and that matures T trading periods from today. The discounted value of this derivative is some function $g(S_0, \dots, S_T, K)e^{-r\delta T}$, where r is the interst rate and δ is the length of a trading period. If the stock price is a random variable, then the fair price of the derivative is $E[g(S_0, \dots, S_T, K)]e^{-r\delta T}$.

It is often assumed that the stock price follows a geometric Brownian motion:

$$(1a) \qquad S_i = S_{i-1} \exp\left[(r - \sigma^2/2)\delta + \sigma \xi_i \sqrt{\delta}\right],$$

where σ is the volatility of the stock, and ξ_i is a standard normal random variable. Then, the fair price of the derivative can be written as

(1b) $$p = E[\tilde{g}(S_0, K, r, \sigma, T, \delta, \xi_1, \ldots, \xi_T)].$$

This expectation is an integral over T-dimensional space, so computing the price of a derivative becomes a problem of multidimensional integration. For high dimensions, product Gauss or Newton-Cotes rules are not effective. Rather, one uses Monte Carlo methods to compute high dimensional integrals because they make weak assumptions about the smoothness of the integrand, and their performance is relatively independent of dimension.

The above example is a simplification. Sometimes derivatives depend on more than one security and/or other parameters. Other computational finance problems, such as value at risk (VaR), do not fit the exact model above. However, in all these cases the principle is the same: the quantity to be computed is an expectation, i.e., is a high dimensional integral.

1.2. Quadrature Error for Monte Carlo Methods. Let the integral that is to be evaluated be written as

$$I(f) \equiv \int_{\mathcal{X}} f(\mathbf{x}) \, dF(\mathbf{x}),$$

where f is some known function, the integration domain, $\mathcal{X}$, is some subset of $\mathbf{R}^s$, and $F(\mathbf{x})$ is some probability distribution with $\int_{\mathcal{X}} dF(\mathbf{x}) = 1$. Monte Carlo quadrature approximates this integral by

(2) $$\hat{I}(f; P) \equiv \frac{1}{n} \sum_{i=1}^{n} f(\mathbf{z}_i),$$

where $P = \{\mathbf{z}_1, \ldots, \mathbf{z}_n\}$ is a set of random numbers. The error of this rule is $\mathrm{Err}(f; P) \equiv I(f) - \hat{I}(f; P)$. Since the error is a random quantity, it does not make sense to look at the error of a particular realization. Rather, one usually looks at the mean square quadrature error:

$$E[I(f) - \hat{I}(f; P)]^2 = \mathrm{Bias}[\hat{I}(f; P)] + \mathrm{Var}[\hat{I}(f; P)],$$

where

$$\mathrm{Bias}[\hat{I}(f; P)] = [I(f) - E\hat{I}(f; P)]^2, \qquad \mathrm{Var}[\hat{I}(f; P)] = E[\hat{I}(f; P) - E\hat{I}(f; P)]^2.$$

This error is divided into two parts, the *bias* and the *variance*. Often unbiased rules are chosen, i.e., $\mathrm{Bias}[\hat{I}(f; P)] = 0$.

For simple Monte Carlo methods, the sample $P = P_{\mathrm{mc}}$ is chosen to be a set of independent and identically distributed points with distribution F. For this unbiased rule the mean square error is

$$E[\mathrm{Err}(f; P_{\mathrm{mc}})]^2 = \mathrm{Var}[\hat{I}(f; P_{\mathrm{mc}})] = \frac{1}{n} \mathrm{Var}(f),$$

where the variance of the integrand is defined as:

$$\mathrm{Var}(f) = I(f^2) - [I(f)]^2.$$

A probabilistic error estimate for Monte Carlo quadrature based on the previous two equations can be written as follows:

$$[\mathrm{Err}(f; P_{\mathrm{mc}})]^2 \approx \frac{1}{n} \left\{ \hat{I}(f^2; P_{\mathrm{mc}}) - [\hat{I}(f; P_{\mathrm{mc}})]^2 \right\}.$$

Provided that the variance of the integrand is bounded, Monte Carlo quadrature error is independent of dimension. More sophisticated Monte Carlo methods, such as stratified sampling, importance sampling, etc., aim to reduce the variance of $\hat{I}$. However, these improvements still give $\text{Var}[\hat{I}(f; P)] = \text{O}(n^{-1})$, i.e., the integration error is $\text{O}(n^{-1/2})$.

2. The Quasi-Monte Carlo Alternative

Quasi-Monte Carlo methods provide a way to improve the convergence rate. These methods have been developed for several decades, but are still an active area of research. For an overview of quasi-Monte Carlo methods see the monographs [**HW81, Nie92, SJ94, Tez95, Fox99**] and the review articles [**PT89, Caf98, Mor98**]. The latest developments in quasi-Monte Carlo methods are reported in the biennial conference proceedings [**NS95, NHLZ98, NS00**].

Consider quadrature rules of the form (2), but now allow P to be chosen deterministically. Note that the quadrature error may be written as

$$\text{Err}(f; P) = I(f) - \hat{I}(f; P) = \int_{\mathcal{X}} f(\mathbf{x})\, d\mathbf{x} - \frac{1}{n} \sum_{i=1}^{n} f(\mathbf{z}_i) = \int_{\mathcal{X}} f(\mathbf{x})\, d[F(\mathbf{x}) - F_P(\mathbf{x})],$$

where

$$F_P(\mathbf{x}) = \frac{1}{n} \sum_{i=1}^{n} 1_{[\mathbf{z}_i, +\infty)}(\mathbf{x})$$

is the *empirical distribution function* for P. Here, $1_{\{\cdot\}}$ denotes the indicator functions, so $1_{[\mathbf{z}_i, +\infty)}(\mathbf{x}) = 1$ if all components of $\mathbf{x}$ are no less than the respective components of the sample point $\mathbf{z}_i$, and $1_{[\mathbf{z}_i, +\infty)}(\mathbf{x}) = 0$ otherwise. If the integrands lie in a Banach space $\mathcal{W}$, then the quadrature error is bounded as:

$$|\text{Err}(f; P)| = \left| \int_{\mathcal{X}} f(\mathbf{x})\, d[F(\mathbf{x}) - F_P(\mathbf{x})] \right| \leq D(P) V(f).$$

Here $V(f)$ is called the *variation* of f and is defined as the non-constant part of f:

$$V(f) = \min_{g \in \mathcal{W}} \{ \|g\|_{\mathcal{W}} : g \in \mathcal{W} \text{ and } f - g = \text{constant} \}.$$

The quantity $D(P)$ is called the *discrepancy* of P, and is defined as the norm of the error functional. The discrepancy can also be interpreted as a *goodness-of-fit statistic* [**Nie92, Hic99, Hic00**]:

$$D(P) = \|F - F_P\|_{\mathcal{M}}, \quad \text{where} \quad \|G(x)\|_{\mathcal{M}} = \sup_{f \neq 0} \frac{\left| \int_{\mathcal{X}} f(x)\, dG(x) \right|}{\|f\|_{\mathcal{W}}}.$$

Here $\|\cdot\|_{\mathcal{M}}$ is a the norm on a space of signed measures.

Quasi-Monte Carlo methods use deterministic point sets, P, with low discrepancy, instead of simple random sets. Low discrepancy sets are also called superuniform point sets. Quasi-Monte Carlo quadrature error is nearly $\text{O}(n^{-1})$ [**Nie92**], and sometimes even better.

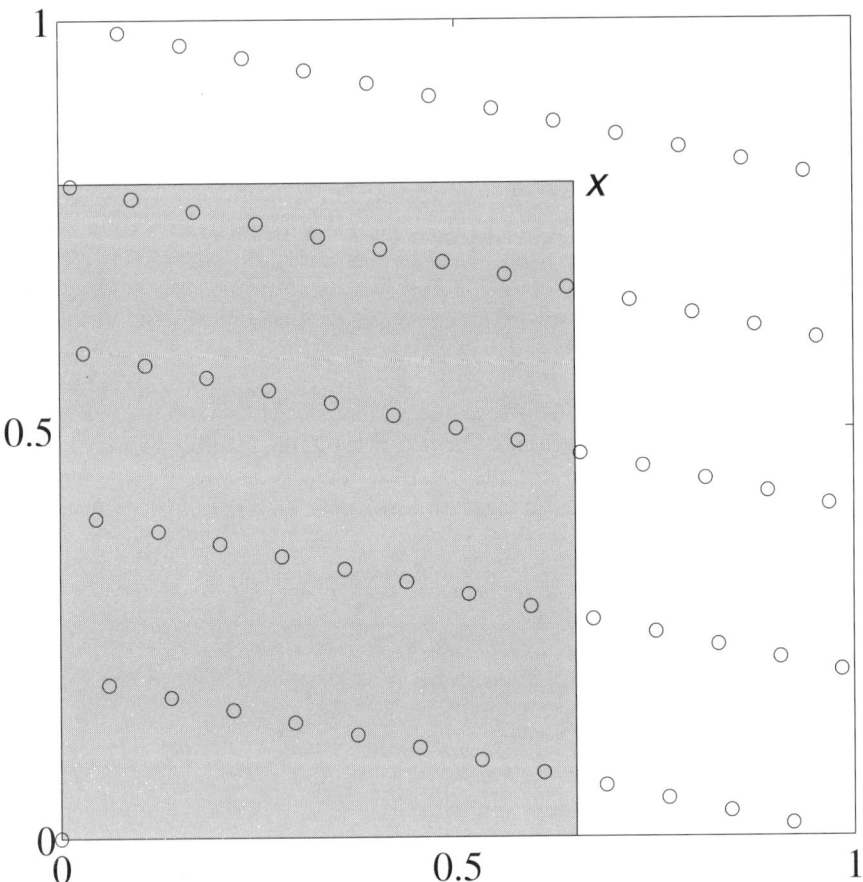

FIGURE 1. The star discrepancy depends on the volume of the box, $F_{\mathrm{unif}}(\mathbf{x})$, and the proportion of points in the box, $F_P(\mathbf{x})$.

2.1. Discrepancies. There are many different discrepancies. Each uniquely defined by the choice of the space of integrands, $\mathcal{W}$, and the norm on that space. Some typical examples are the $\mathcal{L}_\infty$-discrepancy (also known as the Kolmogorov-Smirnov statistic) and the $\mathcal{L}_2$-star discrepancy (also known as the Cramér-von Mises statistic). These are special cases of the $\mathcal{L}_p$-star discrepancy:

$$D_p^*(P) = \|F_{\mathrm{unif}} - F_P\|_p,$$

where F_{unif} denotes the uniform distribution the domain, $\mathcal{X} = [0,1]^s$, and $\|\ \|_p$ denotes the $\mathcal{L}_p$ norm. As shown in Figure 1, the quantity $F_{\mathrm{unif}}(\mathbf{x}) - F_P(\mathbf{x})$ is the volume of the box $[0, \mathbf{x}]$ minus the proportion of points in P that are also in the box $[0, \mathbf{x}]$.

For general Banach spaces of integrands it is difficult to give an explicit formula for the corresponding discrepancy of a set of points. However, if the space of integrands, $\mathcal{W}$, is a Hilbert space of sufficiently smooth functions, then this Hilbert space has a reproducing kernel $K(\mathbf{x}, \mathbf{y})$, such that

$$f(\mathbf{y}) = \langle f, K(\mathbf{x}, \mathbf{y}) \rangle \quad \text{for all } f \in \mathcal{W}, \ \mathbf{y} \in \mathcal{X}.$$

Once this kernel is identified, then the discrepancy of $P = \{\mathbf{z}_1, \ldots, \mathbf{z}_n\}$ can be written immediately as [**Hic00**]:

$$(3a) \qquad [D(P)]^2 = \int_{\mathcal{X}^2} K(\mathbf{x}, \mathbf{y}) \, d[F(\mathbf{x}) - F_P(\mathbf{x})] \times d[F(\mathbf{y}) - F_P(\mathbf{y})]$$

$$(3b) \qquad = \int_{\mathcal{X}^2} K(\mathbf{x}, \mathbf{y}) \, dF(\mathbf{x}) \, dF(\mathbf{y}) - \frac{2}{n} \sum_{i=1}^{n} \int_{\mathcal{X}} K(\mathbf{z}_i, \mathbf{y}) \, dF(\mathbf{y})$$

$$+ \frac{1}{n^2} \sum_{i,k=1}^{n} K(\mathbf{z}_i, \mathbf{z}_k).$$

This formula takes at most $O(n^2)$ operations to evaluate.

Several examples of Hilbert spaces of integrands and the corresponding definitions of discrepancy and variation are given in [**Hic96b, Hic98a, Hic98b, Hic99, Hic00, HH99**]. Some of these discrepancies have geometric interpretations, such as the star discrepancy. An important example of a weighted $\mathcal{L}_2$-star discrepancy comes from [**SW98**]. The reproducing kernel is:

$$(4a) \qquad K(\mathbf{x}, \mathbf{y}) = \prod_{j=1}^{s} \{1 + \beta_j [1 - \max(x_j, y_j)]\},$$

where the β_j are arbitrary positive constants. The corresponding discrepancy for this kernel follows by substituting this kernel into formula (3):

$$(4b) \qquad [D_2^*(P)]^2 = \prod_{j=1}^{s} \left\{1 + \frac{\beta_j}{3}\right\} - \frac{2}{n} \sum_{i=1}^{n} \prod_{j=1}^{s} \left\{1 + \beta_j \left[\frac{1 - z_j^2}{2}\right]\right\}$$

$$+ \frac{1}{n^2} \sum_{i,k=1}^{n} \prod_{j=1}^{s} \{1 + \beta_j [1 - \max(z_j, z_j')]\}$$

$$(4c) \qquad = \sum_{\emptyset \subset u \subseteq \{1:s\}} \left[\left(\prod_{j \in u} \beta_j\right) \|F_{\text{unif},u} - F_{P_u}\|_2^2\right].$$

Here, u denotes a subset of the coordinate indices, and $|u|$ denotes its cardinality. The set P_u is the projection of the set P into the $|u|$-dimensional unit cube, $[0,1]^u$, and $F_{\text{unif},u}$ is the uniform distriubtion on $[0,1]^u$. The corresponding variation for $\mathcal{W}_2^*$, the Hilbert space with the above reproducing kernel, is

$$(5) \qquad [V_2^*(f)]^2 = \sum_{\emptyset \subset u \subseteq \{1:s\}} \left[\left(\prod_{j \in u} \beta_j^{-1}\right) \left\|\left.\frac{\partial^{|u|} f}{\partial x_u}\right|_{x_{\{1:s\}-u}=1}\right\|_2^2\right].$$

Thus, integrands in $\mathcal{W}_2^*$ must have square-integrable mixed partial derivatives.

2.2. Low Discrepancy Sequences. Given a measure of discrepancy it is a nontrivial task to find a low discrepancy set. An even more difficult task is to find an infinite sequence of points whose first n terms is a low discrepancy set for any n. Two main families of low discrepancy sequences are *lattices* and *digital sequences*.

The formula for the node points of a shifted rank-1 lattice is given by

$$(6) \qquad P^{\text{lat}} = \{i\mathbf{h}/n + \mathbf{\Delta} \mod 1 : i = 0, \ldots, n-1\},$$

where $\mathbf{h}$ is an s-dimensional generating vector, and $\mathbf{\Delta}$ is an s-dimensional shift vector. A plot of a particular lattice is given in Figure 2. Discussions of the

FIGURE 2. A shifted lattice.

properties of lattices are given in [**Nie92, SJ94, Hic98b**]. The formula for lattices is rather simple. The difficulty is to find a good generating vector, $\mathbf{h}$, that makes the lattice have low discrepancy for the chosen n and s. Efforts to search for good generating vectors are reported in [**SJ94, L'É99**].

Recently the formula for lattices was extended to an infinite sequence. This is done by using the radical inverse function, $\phi_b(i)$. For any integer $b \geq 2$, let any non-negative integer i be represented in base b as $i = \cdots i_3 i_2 i_1$, where the digits i_k take on values between 0 and $b - 1$. The function $\phi_b(i)$ flips the digits about the decimal point, i.e.,

$$\phi_b(i) = 0.i_1 i_2 i_3 \cdots \text{ base } b = \sum_{i=k}^{\infty} i_k b^{-k}.$$

The sequence $\{\phi_b(i) : i = 0, 1, \dots\}$ is called the *van der Corput sequence*. An infinite sequence of imbedded lattices is defined by replacing i/n by $\phi_b(i)$ in (6):

$$P^{\text{lat-seq}} = \{\phi_b(i)\mathbf{h} + \mathbf{\Delta} \mod 1 : i = 0, 1, \dots\}.$$

The first n points of this sequence are a lattice if n is a power of b. The problem of finding a good generating vector now becomes more difficult because one vector $\mathbf{h}$ must serve for a sequence of n. Tables of generating vectors are given in [**HHLL00**].

Another popular construction of low discrepancy sets is digital sequences. Let $b \geq 2$ be an integer as before, and for simplicity assume that b is prime. Given the generator matrices $\mathbf{C}_1, \ldots, \mathbf{C}_s$ and any non-negative integer $i = \cdots i_3 i_2 i_1$ (base b), one can define the integers $z_{ij} = \cdots z_{ij3} z_{ij2} z_{ij1}$ (base b) by the formula

$$\begin{pmatrix} z_{ij1} \\ z_{ij2} \\ z_{ij3} \\ \vdots \end{pmatrix} = \mathbf{C}_j \begin{pmatrix} i_1 \\ i_2 \\ i_3 \\ \vdots \end{pmatrix} \quad \text{mod } b.$$

Then a digital sequence is defined as

$$P^{\text{dig}-\text{seq}} = \{(\phi_b(z_{i1}), \phi_b(z_{i2}), \ldots, \phi_b(z_{is})) : i = 0, 1, \ldots\}.$$

A digital net is the first b^m points of a digital sequence. Different choices of generator matrices for digital sequences are given in [**Sob67, Fau82, Nie92, NX96**]. One often takes $\mathbf{C}_1$ to be the identity matrix, in which case the first coordinate of the digital sequence is just the van der Corput sequence.

Digital nets and sequences are special cases of (t, m, s)-nets and (t, s)-sequences. A (t, m, s)-net in base b has b^m points. Choose any non-negative integers $k_1, \ldots, k_s$, with $k_1 + \cdots + k_s = m - t$. Then tile the unit cube, $[0, 1)^s$ with b^{m-t} boxes of dimensions $b^{-k_1} \times \cdots \times b^{-k_s}$. Every box must contain b^t points of the (t, m, s)-net, no matter how the k_j are chosen. Smaller values of t imply better nets. The definition of a (t, s)-sequence requires the points numbered $\ell b^m, \ldots, (\ell + 1) b^m$ to form a (t, m, s)-net for any non-negative integer ℓ.

Figure 3 shows a slightly shifted digital $(0, 3, 2)$-net in base 3, which is the first $3^3 = 27$ points of a $(0, 2)$-sequence due to Faure [**Fau82**]. The generating matrices are

$$\mathbf{C}_1 = \mathbf{I}, \quad \mathbf{C}_2 = \begin{pmatrix} 1 & 1 & 1 & \cdots \\ 0 & 1 & 2 & \cdots \\ 0 & 0 & 1 & \cdots \\ \vdots & \vdots & \vdots & \end{pmatrix}.$$

The matrix $\mathbf{C}_2$ is simply a matrix whose upper triangular part contains the Pascal's triangle modulo 3. Note that the tiles in Figure 3 of area $1/27$ each contain one point.

How do the discrepancies of the above points compare to those of a simple random sample? If n is a prime power, then there exist shifted lattices whose weighted $\mathcal{L}_2$-star discrepancy as defined in (4) is $\mathrm{O}(n^{-1}[\log n]^{s-1})$ (see [**Nie92**, Theorem 5.14] and [**Hic98b**]). Moreover, it is also known that there exist (t, m, s)-nets whose weighted $\mathcal{L}_2$-star discrepancy is $\mathrm{O}(n^{-1}[\log n]^{(s-1)/2})$ [**Hic96a, HY00**]. This is the best possible asymptotic order for any set [**Rot54**]. There also exist (t, s)-sequences whose first n points have a weighted $\mathcal{L}_2$-star discrepancy of $\mathrm{O}(n^{-1}[\log n]^{s/2})$ for any n [**HY00**]. In contrast, the root mean square discrepancy of a simple random set is proportional to $n^{-1/2}$. Elementary calculations applied to (3) give

$$E[D^2(P_{\text{mc}})] = \frac{1}{n}\left[\int_{\mathcal{X}} K(\mathbf{x}, \mathbf{x}) \, dF(\mathbf{x}) - \int_{\mathcal{X}^2} K(\mathbf{x}, \mathbf{y}) \, dF(\mathbf{x}) \, dF(\mathbf{y})\right].$$

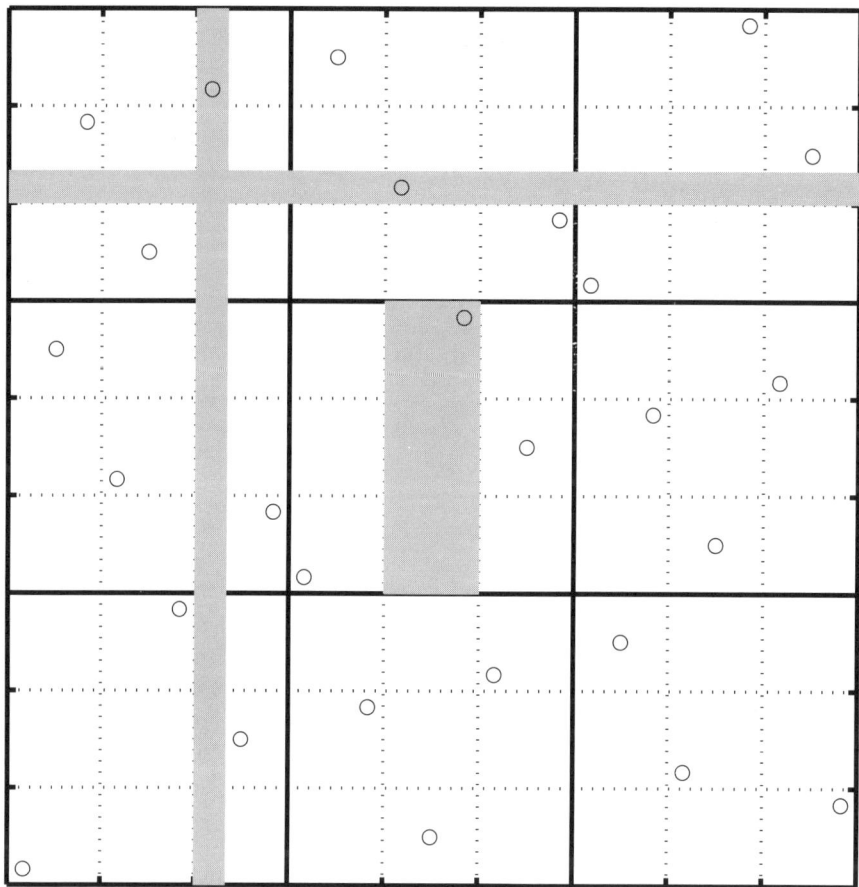

FIGURE 3. A $(0, 3, 2)$-net in base 3.

3. A Monte Carlo Counter-Proposal

As shown in the previous section quasi-Monte Carlo methods can obtain a
better convergence rate than Monte Carlo methods because the points are chosen to be more uniform, i.e., have lower discrepancy. However, in comparison
to Monte Carlo methods, deterministic quasi-Monte Carlo methods have two disadvantages. Because they are deterministic, quasi-Monte Carlo methods are biased. The mean of the quadrature estimate is not the desired integral. Moreover,
whereas Monte Carlo methods have straightforward probabilistic error estimates,
quasi-Monte Carlo methods do not have any simple error estimates. Therefore,
some scholars have suggested randomizing quasi-Monte Carlo sets in a way that
still preserves their low discrepancy. One may think of randomized quasi-Monte
Carlo as a sophisticated sampling technique.

Below, two randomizations of points on the unit cube are described. In both
cases one starts with a low discrepancy set P and obtains a random, low discrepancy
set, P_{ran}, which is used to evaluate the integral according to the rule $\hat{I}(f; P_{ran})$ as
defined in (2).

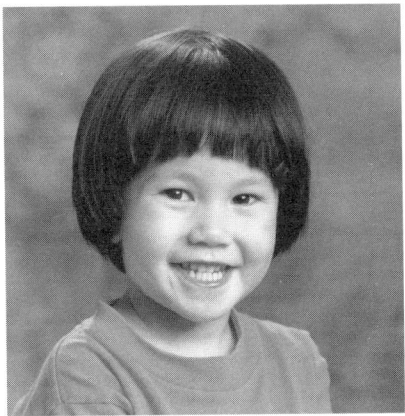

FIGURE 4. A square and its shifted counterpart.

3.1. Two Types of Randomizations. One simple randomization adds the same s-dimensional random shift to every point. That is, if P is the original point set, then $P_{\mathrm{sh}} = \{\mathbf{z} + \mathbf{\Delta} \mod 1 : \mathbf{z} \in P\}$, where $\mathbf{\Delta}$ is a random vector uniformly distributed on $[0,1)^s$. Figure 4 shows this kind of randomization pictorially. This kind of randomization, which was introduced in [**CP76**], is often used with lattice rules because shifted lattice rules retain their lattice structure. However, shifted nets do not necessarily remain nets.

A more sophisticated scrambling was proposed by Art Owen [**Owe95**]. Figure 5 depicts a base 3 version of this scrambling. Starting with the first coordinate, one slices the cube into b equal pieces and randomly permutes them. Then the same operation is applied to each of these pieces. Each big piece is itself split into b smaller pieces and these smaller are randomly permuted, with the random permutations being independent of each other and of the original permutation. This process is carried out recursively as many times as you wish. Then one starts over on the second coordinate and does the same. After that the third, fourth, etc. coordinates are scrambled as well, each independently of the others. Scrambled (t, m, s)-nets are still (t, m, s)-nets with probability one. However, scrambled lattices lose their lattice structure.

3.2. Error of Randomized Quasi-Monte Carlo. The error of randomized quasi-Monte Carlo algorithms using the random set P_{ran} may be analyzed in two ways. If one follows the usual analysis of Monte Carlo algorithms, then one computes the worst possible mean square error. For an unbiased rule this is the worst possible variance of the quadrature error:

$$(7) \qquad \sup_{f \in \mathcal{W}, V(f) \leq 1} \left\{ E[\mathrm{Err}(f; P_{\mathrm{ran}})]^2 \right\} = \sup_{f \in \mathcal{W}, V(f) \leq 1} \mathrm{Var}[\hat{I}(f; P_{\mathrm{ran}})].$$

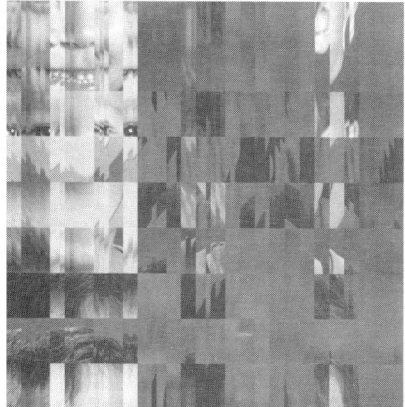 

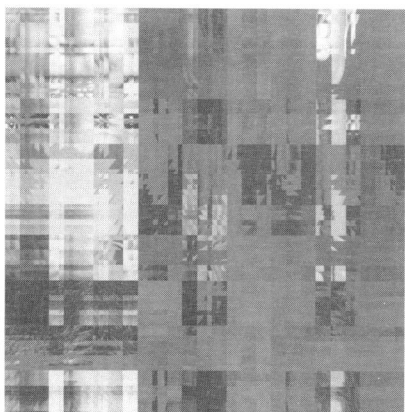

FIGURE 5. A square and its scrambled counterpart in steps.

If one follows the usual analysis of quasi-Monte Carlo algorithms, then one computes the mean square of the worst possible error, which is the mean square discrepancy:

$$(8) \qquad E\left\{\sup_{f\in\mathcal{W}, V(f)\leq 1}[\mathrm{Err}(f; P_{\mathrm{ran}})]^2\right\} = E\left\{D^2(P_{\mathrm{ran}}; K)\right\}.$$

For the shift and scramble randomizations described above, and perhaps for other randomizations as well, the mean square discrepancy can be calculated more easily by finding a *filtered* kernel, K_{ran}, depending on K, such that

$$E\left\{D^2(P_{\mathrm{ran}}; K)\right\} = D^2(P; K_{\mathrm{ran}}),$$

where P is the original set before randomization. For the definitions of the filtered kernels for shifting and scrambling see [**Hic98b, HY00, HW01**].

The criterion in (7) corresponds to the case in which your enemy picks the worst integrand possible *before* you choose your random quadrature rule. The criterion in (8) corresponds to the case in which your enemy picks the worst integrand possible *after* you choose your random rule. Thus, the former criterion is always less than or equal to the latter. In fact, it is shown in [**HW01**] that (7) corresponds to the spectral radius of an infinite dimensional matrix, while (8) corresponds to the trace of the same matrix.

For good randomly shifted lattices, $P_{\text{sh}}^{\text{lat}}$, the root mean square weighted $\mathcal{L}_2$-star discrepancy, as defined in (4), has the same asymptotic order as described at the end of Section 2 (see [**Nie92**, Theorem 5.14] and [**Hic98b**]):

$$\sqrt{E[D_2^*(P_{\text{sh}}^{\text{lat}})]^2} = \mathrm{O}(n^{-1}[\log n]^{s-1}).$$

In fact, the result quoted at the end of Section 2 relies on taking the average over random shifts. If the discrepancy (3) is defined using a reproducing kernel with sufficient *smoothness and periodicity*, then the root mean square discrepancy is

$$\sqrt{E[D(P_{\text{sh}}^{\text{lat}}; K)]^2} = \mathrm{O}(n^{-\alpha}[\log n]^{\alpha(s-1)}), \quad \alpha > 1.$$

Although the convergence rate looks very attractive as α gets large, most integrands in practice do not satisfy the periodicity required to obtain this convergence rate. One may transform the integrand to periodize it, but such transformations tend to amplify the variation of the integrand, thus nullifying the benefits of periodization.

The quadrature errors of scrambled (t, m, s)-nets, $P_{\text{scr}}^{\text{net}}$, and scrambled (t, s)-sequences, $P_{\text{scr}}^{\text{seq}}$, have been studied by [**Owe95, Hic96b, Owe97a, Owe97b, HH99, Yue99, YM99, HY00**]. Under quite general conditions nets are not much worse than Monte Carlo:

$$\sqrt{\sup_{V(f) \leq 1} \mathrm{Var}[\hat{I}(f; P_{\text{scr}}^{\text{net}})]} \leq c \sqrt{\frac{1}{n} \mathrm{Var}(f)},$$

$$\sqrt{E[D(P_{\text{scr}}^{\text{net}}; K)]^2} \leq c \sqrt{E[D(P_{\text{mc}}; K)]^2}.$$

A similar result holds for scrambled sequences. For mildly smooth integrands, i.e., those with finite variation as defined in (5), nets have higher order convergence:

$$\left. \begin{array}{l} \sqrt{\sup_{V(f) \leq 1} \mathrm{Var}[\hat{I}(f; P_{\text{scr}}^{\text{net}})]} \\[2mm] \sqrt{E_{\text{scr}}[D_2^*(P_{\text{scr}}^{\text{net}})]^2} \end{array} \right\} = \mathrm{O}(n^{-1}[\log n]^{(s-1)/2}).$$

For integrands with greater smoothness, namely square-integrable second order mixed partial derivatives, even higher order convergence is possible:

$$(9) \qquad \left. \begin{array}{l} \sqrt{\sup_{V(f) \leq 1} \mathrm{Var}[\hat{I}(f; P_{\text{scr}}^{\text{net}})]} \\[2mm] \sqrt{E_{\text{scr}}[D(P_{\text{scr}}^{\text{net}}; K)]^2} \end{array} \right\} = \mathrm{O}(n^{-3/2}[\log n]^{(s-1)/2}).$$

However, if one looks at the first n points of a (t, s)-sequence, with mild or greater smoothness, one only gets:

$$\left. \begin{array}{l} \sqrt{\sup_{V(f) \leq 1} \mathrm{Var}[\hat{I}(f; P_{\text{scr}}^{\text{seq}})]} \\[2mm] \sqrt{E_{\text{scr}}[D(P_{\text{scr}}^{\text{seq}}; K)]^2} \end{array} \right\} = \mathrm{O}(n^{-1}[\log n]^{s/2}).$$

Result (9) means that for smooth enough integrands, there exist good nets that give an error of $\mathrm{O}(n^{-3/2+\epsilon})$. However, it is not yet known how to construct them directly. The best one can do is to take a known net, which may not have superior error properties, and randomly scramble it. The resulting net is likely to give $\mathrm{O}(n^{-3/2+\epsilon})$ error.

4. Dimension: Good and Bad News

In actual numerical computations of the discrepancy for low discrepancy sets one notices that the discrepancy decays like $O(n^{-1/2})$ for small n before obtaining its asymptotic behavior [**MC94, Hic95**]. The value of n required to reach the asymptotic behavior of $O(n^{-1+\epsilon})$ increases with dimension, s. From another point of view, if the dimension, s, becomes arbitrarily large, then the term $[\log n]^{(s-1)/2}$ appearing above can be enormous. Thus, one might expect that quasi-Monte Carlo methods would not work well for finance problems, where the dimension can be large.

The dimension of an integrand can be tricky thing to find. Consider the function

(10)
$$f(x_1, x_2, x_3, x_4) = x_1 + x_2 \sin(x_3).$$

What is the dimension? At least three answers are possible. The nominal dimension is 4. However, this function only depends on the first three variables. Moreover, this function is a sum of pieces, each of which only depend on two variables. So, the effective dimension of this function could be said to be 2 or 3, depending on one's definition. Precise definitions of the effective dimension are given by Caflisch, Morokoff and Owen [**CMO97**].

One may write an integrand, f, as a sum of parts, f_u, each of which depends only on some of the coordinates, $\mathbf{x}_u$. For example, for functions with finite variation defined by (5) one could define iteratively

$$f_u(\mathbf{x}_u) = f(\mathbf{x}_u, \mathbf{1}) - \sum_{v \subset u} f_v(\mathbf{x}_v), \quad \emptyset \subseteq u \subseteq \{1, \dots, s\}.$$

For example, the parts of the above function (10) are:

$$f_\emptyset = 1 + \sin(1), \quad f_{\{1\}} = x_1 - 1, \quad f_{\{2\}} = \sin(1)(x_2 - 1),$$
$$f_{\{3\}} = \sin(x_3) - \sin(1), \quad f_{\{2,3\}} = (x_2 - 1)(\sin(x_3) - \sin(1)),$$
$$f_{\{4\}} = f_{\{1,2\}} = f_{\{1,3\}} = f_{\{1,4\}} = f_{\{2,3\}} = f_{\{2,4\}} = f_{\{3,4\}} = 0,$$
$$f_{\{1,2,3\}} = f_{\{1,2,4\}} = f_{\{1,3,4\}} = f_{\{2,3,4\}} = f_{\{1,2,3,4\}} = 0.$$

Since $f = \sum_u f_u$, it follows that $\mathrm{Err}(f; P) = \sum_u \mathrm{Err}(f_u; P)$. The error of each part can be bounded as

$$|\mathrm{Err}(f_u; P)| \leq \|F_{\mathrm{unif},u} - F_{P_u}\|_2 \left\| \frac{\partial^{|u|} f_u}{\partial x_u} \right\|_2.$$

Thus, the error of a part of the integrand is small if *either* the points in the projection of P into the cube $[0, 1)^u$ have low discrepancy, or the norm of the mixed partial derivative of f_u is small. Therefore, for a function with large nominal dimension, s, but small effective dimension (small parts f_u with $|u|$ large), the requirements on uniformity of the points P are not so stringent.

Sloan, Woźniakowski, and Hickernell [**SW98, HW00**] have made this precise. If the weights β_j in the definition of the weighted $\mathcal{L}_2$-star discrepancy satisfy $\sum_{j=1}^{\infty} \beta_j^{1/2} < \infty$, then there exists good sets P with

$$D_2^*(P) = O(n^{-1+\epsilon})$$

uniformly in s. This result means that the asymptotic convergence rate for arbitrary high dimensions is essentially the same as for dimension $s = 1$. This condition on the β_j means that the higher numbered coordinates have less effect on the value of the integrand.

Caflisch, Morokoff and Owen [CMO97] showed that some finance problems have low effective dimension. They also showed how using a Brownian bridge can lower the effective dimension of finance problems. Acworth, Broadie and Glasserman [ABG98] have used principal components analysis to lower the effective dimension in finance problems.

5. Numerical Example

To illustrate the theory described in the previous sections, Monte Carlo and randomized quasi-Monte Carlo methods are applied to the example of an arithmetic mean call option. Suppose that the fair price of the option in (1), i.e., the expected discounted value of the payoff function, is

$$(11) \qquad p = E\left[e^{-r\delta T}\max\left(\frac{1}{T}(S_1 + S_2 + \cdots + S_T) - K, 0\right)\right]$$

Thus, the option holder exercises the option if the arithmetic mean stock price is above the strike price. This expected value above may be written as

$$(12\text{a}) \qquad p = \int_{(-\infty,\infty)^T} P(\xi_1, \ldots, \xi_T)\Phi'(\xi_1)\cdots\Phi'(\xi_1)\, d\xi_1 \cdots d\xi_T,$$

where

$$(12\text{b}) \quad P(\xi_1, \ldots, \xi_T) = e^{-r\delta T}\max\left(\frac{1}{T}S_0\left\{\exp\left[(r - \sigma^2/2)\delta + \sigma\xi_1\sqrt{\delta}\right]\right.\right.$$
$$+ \exp\left[(r - \sigma^2/2)(2\delta) + \sigma(\xi_1 + \xi_2)\sqrt{\delta}\right] + \cdots$$
$$\left.\left. + \exp\left[(r - \sigma^2/2)(T\delta) + \sigma(\xi_1 + \cdots + \xi_T)\sqrt{\delta}\right]\right\} - K, 0\right),$$

and Φ denotes the probability distribution of a standard normal random variable. Since virtually all low discrepancy sets are generated on the unit cube, it is convenient to use a change of variables can be used to transform this integral to one over the unit cube. Letting $\xi_j = \alpha\Phi^{-1}(x_j), j = 1, \ldots, T$, it follows that

$$(12\text{c}) \quad p = \int_{[0,1]^T} P(\alpha\Phi^{-1}(x_1), \ldots, \alpha\Phi^{-1}(x_T))$$
$$\times \exp\left(-[\alpha^2 - 1]\{[\Phi^{-1}(x_1)]^2 + \cdots + [\Phi^{-1}(x_T)]^2\}/2\right)\, dx_1 \cdots dx_T,$$

where the parameter α is normally chosen to be larger than one.

The numerical experiments discussed below compute the options price using simple Monte Carlo, randomly shifted integration lattices, and randomly scrambled nets. The number of points (paths) used for each method is $n = 1, 2, \ldots, 2^{17}$. The generating vector for the extensible rank-1 integration lattices is

$$\mathbf{h} = (1, 17797, 17797^2, \ldots, 17797^{T-1}),$$

as suggested by [HHLL00]. The scrambled net is computed from the Sobol' sequence [Sob67], a (t, s)-sequence in base 2, scrambled using the algorithm given in [HH01]. For the Monte Carlo method, the Gaussian random number generator and the quadrature formula (2) are applied to (12a). For the quasi-Monte Carlo methods, quadrature formula is applied to the transformed integrand (12c). For each method the root mean square errors are computed based on 1000 replications.

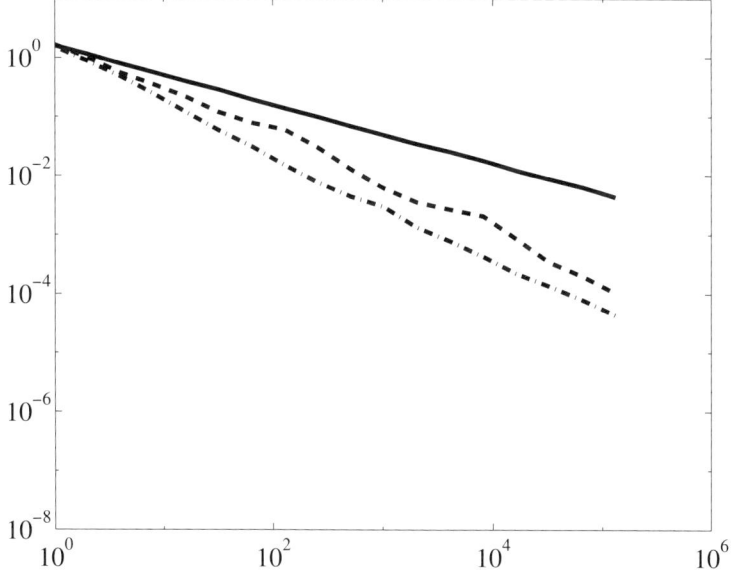

FIGURE 6. Root mean square error for Monte Carlo (solid), shifted lattice rules (dashed) and scrambled nets (dot-dashed) for the case (13).

TABLE 1. Convergence rates q such that the root mean square error is $O(n^{-q})$

σ	0.5	0.01		0.5	
T	4		12	4	
α	1	2		1	
				Control Variates	
Price p	$6.986	$0.997	$0.997	$6.241	$6.986
Monte Carlo	0.50	0.50	0.50	0.50	0.50
Shifted Lattices	0.81	0.91	1.63	0.68	0.69
Scrambled Nets	0.90	1.01	0.97	0.75	0.71

The "exact price" is computed by taking a 1000 replications of a large shifted lattice using control variates (see below).

Suppose that the parameters in (12) are assigned the values

(13a) $S_0 = K = \$100, \quad r = 0.07 \text{ year}^{-1}, \quad \sigma = 0.5 \text{ year}^{-1/2},$

(13b) $T = 4, \quad \delta = 3 \times 7/365 \text{ year} = 3 \text{ weeks},$

(13c) $\delta T = 12 \times 7/365 \text{ year} = 12 \text{ weeks}.$

The price for this option is $p = \$6.986$. Figure 6 shows the root mean square error in computing this option price using simple Monte Carlo, randomly shifted integration lattices, and randomly scrambled nets. The parameter α is set equal to unity.

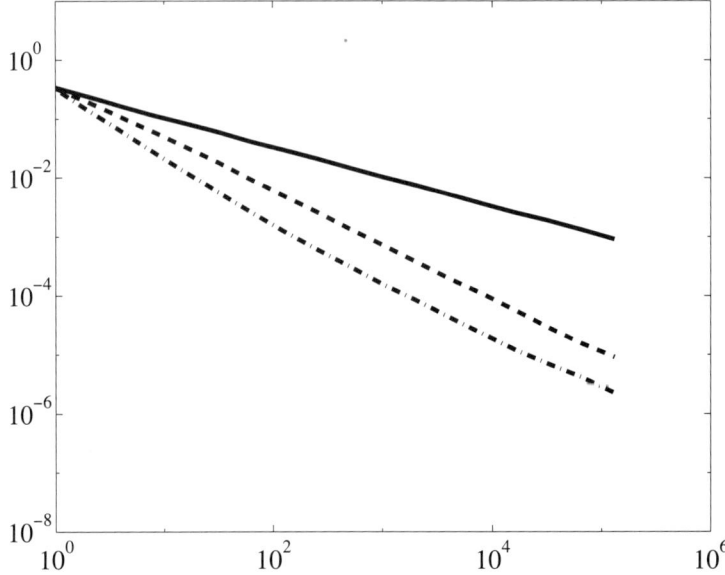

FIGURE 7. Same as Figure 6 but with $\sigma = 0.01$.

The root mean square errors of randomized quasi-Monte Carlo methods are superior to the Monte Carlo method for this example. (Note that the root mean square error for $n = 1$ is the same for all methods, because a shifted one-point lattice and a scrambled one-point net are the same as a one-point Monte Carlo simulation.) The convergence rates of the various methods are recorded in Table 1. The convergence rate for the Monte Carlo method is 0.5, as expected. The convergence rates for the quasi-Monte Carlo methods are better than that of the Monte Carlo method, but do not quite reach 1. This is due in part to the fact that the the n is not large enough for the rates to reach their asymptotic values, and because the integrand is not smooth enough.

If the volatility is lowered, then the effective dimension is lower and the asymptotic rate of decay is reached more quickly. Figure 7 shows the root mean square errors for the same problem as (13) but with a volatility of only $\sigma = 0.01$ year$^{-1/2}$. The convergence rates of the randomized quasi-Monte Carlo methods are more favorable.

For this example, the integrand in (12c) is unbounded on the unit cube for $\alpha = 1$ because it tends to infinity as any x_j tends to one. Increasing the value of α to two has the effect of smoothing the integrand and increases the convergence rate for the randomly shifted lattices (see Figure 8). The convergence rate for the scrambled nets does not improve, perhaps because the value of n is not yet large enough.

If the dimension is increased, e.g., taking $T = 12$ and $\delta = 1$ week, then the convergence rates of the quasi-Monte Carlo methods become worse (see Figure 9). However the convergence rates and the root mean square errors for quasi-Monte Carlo methods are still better than those for Monte Carlo methods.

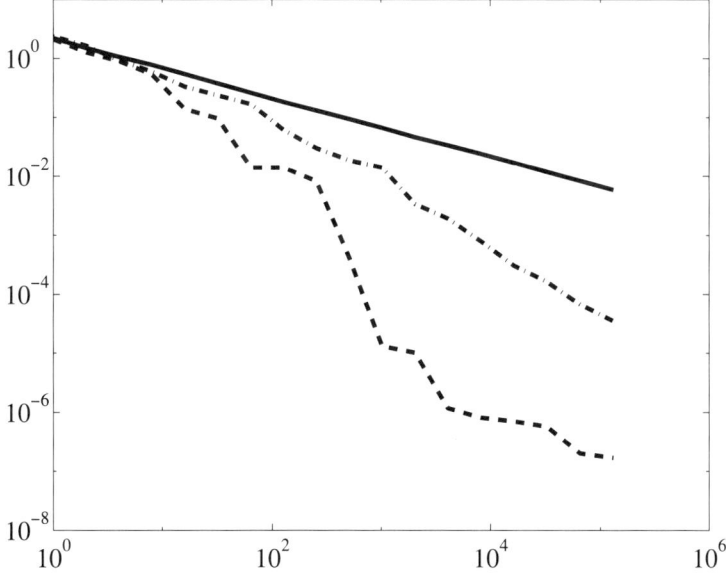

FIGURE 8. Same as Figure 7 but with $\alpha = 2$.

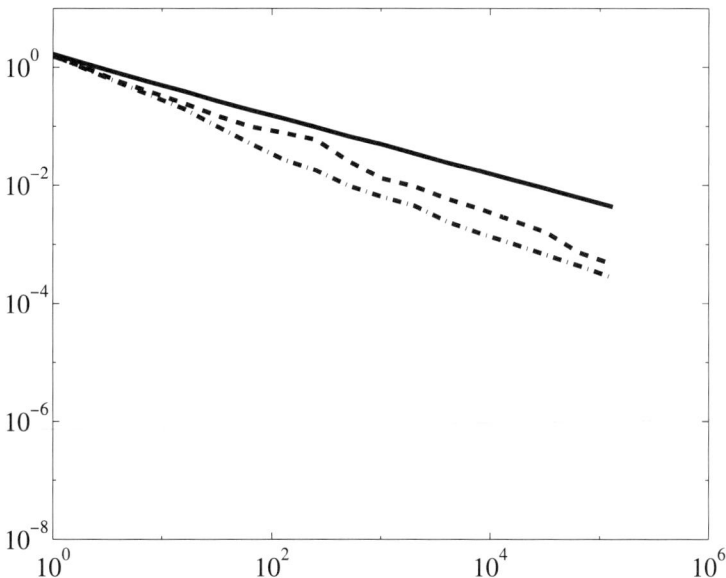

FIGURE 9. Same as Figure 6 but with $T = 12$ and $\delta = 1$ week.

There are several variance reduction methods that can be used for Monte Carlo methods. One of these is control variates. The price of a geometric mean option may be computed exactly, and then simulation can be used to compute the difference between the two prices. Control variates can be used just as easily with Monte Carlo and quasi-Monte Carlo simulation. Using control variates for the example

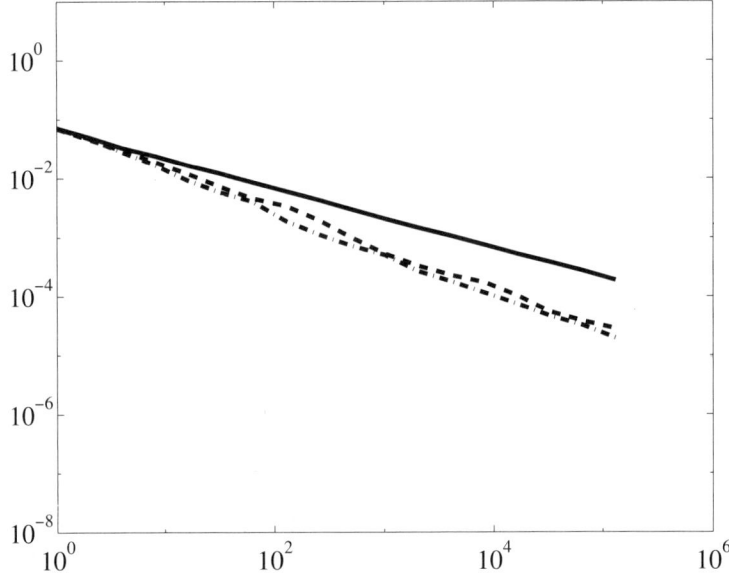

FIGURE 10. Same as Figure 6 but using control variates

(13) decreases the error of both Monte Carlo and quasi-Monte Carlo methods. The convergence rates of the quasi-Monte Carlo methods become somewhat worse, probably because the new integrand, although smaller, is not as smooth as before.

6. Conclusions

Quasi-Monte Carlo methods can improve upon the convergence rate of Monte Carlo methods, reducing the error from $O(n^{-0.5})$ to $O(n^{-1+\epsilon})$. In some cases random shifts of lattices and random scrambling of nets can even improve the convergence rate even further. Randomized quasi-Monte Carlo methods also remove the bias from deterministic quasi-Monte Carlo methods and facilitate probabilistic error estimates.

There are other ways to combine Monte Carlo and quasi-Monte Carlo methods. For very high dimensional problems one may use Latin hypercube sampling [**Owe98**]. One may also use traditional Monte Carlo variance reduction techniques with low discrepancy sequences.

Acknowledgements

The author would like to thank the anonymous referees for their constructive comments. He would also like to thank the organizers of the Worskhop on Applied Probability at Chinese University of Hong Kong for a very interesting meeting. This research was funded by a Hong Kong Baptist University Faculty Research Grant FRG/97-98/II-99.

References

[ABG98]   P. Acworth, M. Broadie, and P. Glasserman, *A comparison of some Monte Carlo techniques for option pricing*, In Niederreiter et al. [**NHLZ98**], pp. 1–18.

[Caf98]   R. E. Caflisch, *Monte Carlo and quasi-Monte Carlo methods*, Acta Numer. **7** (1998), 1–49.

[CMO97] R. E. Caflisch, W. Morokoff, and A. Owen, *Valuation of mortgage backed securities using Brownian bridges to reduce effective dimension*, J. Comput. Finance **1** (1997), 27–46.

[CP76]   R. Cranley and T. N. L. Patterson, *Randomization of number theoretic methods for multiple integration*, SIAM J. Numer. Anal. **13** (1976), 904–914.

[Fau82]   H. Faure, *Discrépance de suites associées à un système de numération (en dimension s)*, Acta Arith. **41** (1982), 337–351.

[Fox99]   B. L. Fox, *Strategies for quasi-Monte Carlo*, Kluwer Academic Publishers, Boston, 1999.

[HH99]   F. J. Hickernell and H. S. Hong, *The asymptotic efficiency of randomized nets for quadrature*, Math. Comp. **68** (1999), 767–791.

[HH01]   H. S. Hong and F. J. Hickernell, *Implementing scrambled digital nets*, 2001, submitted to ACM TOMS.

[HHLL00] F. J. Hickernell, H. S. Hong, P. L'Écuyer, and C. Lemieux, *Extensible lattice sequences for quasi-Monte Carlo quadrature*, SIAM J. Sci. Comput. **22** (2000), 1117–1138.

[Hic95]   F. J. Hickernell, *A comparison of random and quasirandom points for multidimensional quadrature*, In Niederreiter and Shiue [**NS95**], pp. 213–227.

[Hic96a] F. J. Hickernell, *The mean square discrepancy of randomized nets*, ACM Trans. Model. Comput. Simul. **6** (1996), 274–296.

[Hic96b] F. J. Hickernell, *Quadrature error bounds with applications to lattice rules*, SIAM J. Numer. Anal. **33** (1996), 1995–2016, corrected printing of Sections 3-6 in ibid., **34** (1997), 853–866.

[Hic98a] F. J. Hickernell, *A generalized discrepancy and quadrature error bound*, Math. Comp. **67** (1998), 299–322.

[Hic98b] F. J. Hickernell, *Lattice rules: How well do they measure up?*, In Hellekalek and Larcher [**HL98**], pp. 109–166.

[Hic99]   F. J. Hickernell, *Goodness-of-fit statistics, discrepancies and robust designs*, Statist. Probab. Lett. **44** (1999), 73–78.

[Hic00]   F. J. Hickernell, *What affects the accuracy of quasi-Monte Carlo quadrature?*, In Niederreiter and Spanier [**NS00**], pp. 16–55.

[HL98]   P. Hellekalek and G. Larcher (eds.), *Random and quasi-random point sets*, Lecture Notes in Statistics, vol. 138, Springer-Verlag, New York, 1998.

[HW81]   L. K. Hua and Y. Wang, *Applications of number theory to numerical analysis*, Springer-Verlag and Science Press, Berlin and Beijing, 1981.

[HW00]   F. J. Hickernell and H. Woźniakowski, *Integration and approximation in arbitrary dimensions*, Adv. Comput. Math. **12** (2000), 25–58.

[HW01]   F. J. Hickernell and H. Woźniakowski, *The price of pessimism for multidimensional quadrature*, J. Complexity **17** (2001), to appear.

[HY00]   F. J. Hickernell and R. X. Yue, *The mean square discrepancy of scrambled (t, s)-sequences*, SIAM J. Numer. Anal. **38** (2000), 1089–1112.

[L'É99]   P. L'Écuyer, *Tables of linear congruential generators of different sizes and good lattice structure*, Math. Comp. **68** (1999), 249–260.

[LL98]   C. Lemieux and P. L'Écuyer, *Efficiency improvement by lattice rules for pricing asian options*, Proc. 1998 Winter Simulation Conference, IEEE Press, 1998, pp. 579–586.

[MC94]   W. J. Morokoff and R. E. Caflisch, *Quasi-random sequences and their discrepancies*, SIAM J. Sci. Comput. **15** (1994), 1251–1279.

[Mor98]   W. J. Morokoff, *Generating quasi-random paths for stochastic processes*, SIAM Rev. **40** (1998), 765–788.

[NHLZ98] H. Niederreiter, P. Hellekalek, G. Larcher, and P. Zinterhof (eds.), *Monte Carlo and quasi-Monte Carlo methods 1996*, Lecture Notes in Statistics, vol. 127, Springer-Verlag, New York, 1998.

[Nie92]   H. Niederreiter, *Random number generation and quasi-Monte Carlo methods*, CBMS-NSF Regional Conference Series in Applied Mathematics, SIAM, Philadelphia, 1992.

[NS95]   H. Niederreiter and P. J.-S. Shiue (eds.), *Monte Carlo and quasi-Monte Carlo methods in scientific computing*, Lecture Notes in Statistics, vol. 106, Springer-Verlag, New York, 1995.

[NS00]   H. Niederreiter and J. Spanier (eds.), *Monte Carlo and quasi-Monte Carlo methods 1998*, Springer-Verlag, Berlin, 2000.

[NT96]   S. Ninomiya and S. Tezuka, *Toward real-time pricing of complex financial derivatives*, Appl. Math. Finance **3** (1996), 1–20.

[NX96]   H. Niederreiter and C. Xing, *Quasirandom points and global function fields*, Finite Fields and Applications (S. Cohen and H. Niederreiter, eds.), London Math. Society Lecture Note Series, no. 233, Cambridge University Press, 1996, pp. 269–296.

[OT97]   A. B. Owen and D. T. Tavella, *Scrambled nets for value-at-risk calculations*, VAR Understanding and applying value-at-risk (London) (S. Grayling, ed.), Risk Publications, 1997, pp. 257–273.

[Owe95]   A. B. Owen, *Randomly permuted (t, m, s)-nets and (t, s)-sequences*, In Niederreiter and Shiue [**NS95**], pp. 299–317.

[Owe97a] A. B. Owen, *Monte Carlo variance of scrambled net quadrature*, SIAM J. Numer. Anal. **34** (1997), 1884–1910.

[Owe97b] A. B. Owen, *Scrambled net variance for integrals of smooth functions*, Ann. Stat. **25** (1997), 1541–1562.

[Owe98]   A. B. Owen, *Scrambling Sobol' and Niederreiter-Xing points*, J. Complexity **14** (1998), 466–489.

[PT89]   W. H. Press and S. A. Teukolsky, *Quasi- (that is, sub-) random numbers*, Computers in Physics **3** (1989), no. 6, 76–79.

[PT95]   S. Paskov and J. Traub, *Faster valuation of financial derivatives*, J. Portfolio Management **22** (1995), 113–120.

[PT96]   A. Papageorgiou and J. F. Traub, *Beating Monte Carlo*, Risk **9** (1996), no. 6, 63–65.

[Rot54]   K. F. Roth, *On irregularities of distribution*, Mathematika **1** (1954), 73–79.

[SJ94]   I. H. Sloan and S. Joe, *Lattice methods for multiple integration*, Oxford University Press, Oxford, 1994.

[Sob67]   I. M. Sobol', *The distribution of points in a cube and the approximate evaluation of integrals*, U.S.S.R. Comput. Math. and Math. Phys. **7** (1967), 86–112.

[SW98]   I. H. Sloan and H. Woźniakowski, *When are quasi-Monte Carlo algorithms efficient for high dimensional integrals*, J. Complexity **14** (1998), 1–33.

[Tez95]   S. Tezuka, *Uniform random numbers: Theory and practice*, Kluwer Academic Publishers, Boston, 1995.

[Tez98]   S. Tezuka, *Financial applications of Monte Carlo and quasi-Monte Carlo methods*, In Hellekalek and Larcher [**HL98**], pp. 303–332.

[YM99]   R. X. Yue and S. S. Mao, *On the variance of quadrature over scrambled nets and sequences*, Statist. Probab. Lett. **44** (1999), 267–280.

[Yue99]   R. X. Yue, *Variance of quadrature over scrambled unions of nets*, Statist. Sinica **9** (1999), 451–473.

DEPARTMENT OF MATHEMATICS, HONG KONG BAPTIST UNIVERSITY, FRED@HKBU.EDU.HK, HTTP://WWW.MATH.HKBU.EDU.HK/~FRED

AMS/IP Studies in Advanced Mathematics
Volume 26, 2002

Contingent claim approach for analyzing the creditrisk of defaultable currency swaps

Hong Yu and Yue Kuen Kwok

ABSTRACT. In this paper, we analyze the credit risk associated with defaultable currency swaps under the contingent claim analysis framework. One of the swap parties is subject to intertemporal default risk while the other swap party is assumed to be default free. The event triggering the intertemporal swap default is endogenized by characterizing the creditworthiness of the defaultable swap party. The firm value is used to determine whether the defaultable swap party can fulfil the cashflows associated with the swap contract and other financial obligations. The impact of various clauses and settlement rules in the currency swap contracts are examined. The influences of the rate risk on the swap rates, as exemplified by fluctuating volatilities of the exchange rate, varying correlation between firm value and exchange rate, are also analyzed.

1. INTRODUCTION

In simple terms, a financial swap is the exchange of cashflow or asset based on an underlying index under some prescribed terms in the swap contract. The growth of the swap markets has been phenomenal since the first swap contract structured in 1981. There are a number of explanations for the popularity of swaps. Swaps allow firms to exploit market inefficiencies and to arbitrage quality spread differentials. Also, swaps may be used to adjust the repricing interval of a firm's assets or liabilities in order to reduce the interest rate risk or the exchange rate risk.

Since any swap involves mutual obligations to exchange cashflows, default occurs if a counterparty owes a payment and becomes insolvent. The two major risks of swap contracts are the rate risk and the credit risk. The rate risk arises from the change in the interest rate or the exchange rate, while the credit risk occurs since either swap party may default. Unlike debt contracts, the swap default risk is two-sided since it depends on the realized path of the underlying index of the swap which can result in either party to make payments at different times in the life of the swap.

The maximum loss associated with the credit risk is measured by the swap's replacement cost. The question is: how much spread is appropriate to cover the swap credit risk between the higher-rated and lower-rated firms. Since the credit

Corresponding author Yue Kuen Kwok; e-mail: maykwok@ust.hk, fax number: (852) 2358 1643.

risk borne by swap dealers is far smaller in proportion to the principal than in debt contracts, it is meaningless to estimate the spreads on higher-rated and lower-rated swaps from the qualify spread on bonds of similar credit categories. There are other unique features regarding contract clauses and settlement rules in swap contracts that may affect the credit risk of swaps. For example, the limited feature of a swap discharges all the non-defaulting counterparty's obligations in the event of default, and the settlement can be net or gross, the swaps can be junior, pari-passu or senior than debts and other liabilities.

Hazard rate approach and contingent claim approach

In the literature, there are two popular approaches for pricing the credit spreads of risky financial instruments. These two approaches are commonly called the hazard rate approach and the contingent claim approach (or the firm value approach). The hazard rate models specify exogenously the processes for bankruptcy and the payoff on the risky instruments conditional on default. The model introduces credit classes, with transition from one class to another driven by a deterministic credit migration matrix to account for credit ratings behaviors. Jarrow and Turnbull (1997) and Nielsen and Ronn (1997) applied the hazard rate approach to evaluate the impact of default risk on risky swaps. The hazard rate model views financial instruments as members from a credit rating class, and so it loses sight of the specific features associated with the issuing firm and contractual terms in the instrument. If we would like to investigate the impact of some specified contractual terms and settlement clauses in swap contracts, then it would be more preferable to follow the contingent claim approach (or the firm value approach).

Previous works on default risk analysis of swaps

It has been observed in the financial markets that swap rates seem not to depend sensibly on the credit ratings of the counterparties (Litzenberger, 1992). One of the earlier works on the use of the contingent claim approach in default analysis of swaps is the risk model by Cooper and Mello (1991). Their model assumes a simple capital structure, where one of the swap counterparties is risky and the other is riskless. The firm of the risky party also has debt outstanding. The remainder of the risky firm is financed with equity. Both the debt and swap are assumed to mature on the same date, and the swap is junior in claims than the debt. Cooper and Mellon's model is later extended by Baz and Pascutti (1996) where various swap clauses are considered. They considered swap covenants along three dimensions: settlement of swap amounts can be net or gross, seniority regimes for swaps relative to debt can be senior, pari-passu or junior, and counterparty obligations under insolvency. Li (1998) developed a more refined firm value model where swaps are valued as contingent claims which pay a stream of cash flow equal to the net payments between the swap counterparties.

In the above swap default analysis models, the reorganization of the counterparties when their firm values fall below some threshold values and possible deviations from the strict priority rules have been neglected. At the next level of sophistication of credit risk models, the defaultable swap models should allow intertemporal default risk of the counterparties prior maturity of the swap contract.

Scope of the paper

In this paper, the contingent claim approach is used to construct an equilibrium model for analyzing default risk of currency swaps, where the sources of uncertainty in the models are the rate distribution of the exchange rate, and the firm value of one of the swap parties. For simplicity of analysis, we assume that only one of the swap parties is defaultable. The default risk and the exchange rate risk, and their interaction are analyzed in a combined framework in the contingent claim models, rather than being artificially decomposed as in other approaches. The success of the contingent claim analysis method relies on the precise prescription of the processes that lead to financial distress of the swap parties and the bankruptcy terms upon default. The description of these default conditions are translated into the auxiliary conditions of the contingent claim models. The pricing behaviors of currency swaps with intertemporal default possibilities are examined critically. In particular, special considerations are directed to the swap rate spreads with respect to different parameter values. The set of parameters include the variance rates of the firm value process and the exchange rate process, the correlation coefficient between the processes, and the threshold level of default.

2. FORMULATION OF DEFAULTABLE CURRENCY SWAP MODELS

The option pricing theory has been shown to provide a universal valuation framework for contingent claims. The contingent claim approach on the risk analysis of corporate debts, initiated by Merton (1974), represents an elegant use of the option pricing theory in corporate finance, where corporate liabilities can be viewed as combinations of option contracts. The various claims (such as debt and swap contracts) are modelled in terms of appropriate auxiliary conditions in the contingent claim models. The riskiness of a financial instrument is measured by its risk premium, which is the spread between the prices of the risky instrument and its non-risky counterparty.

Later enhancements of the contingent claim analysis of credit risk include the possibilities of early default, where the firm defaults when the value of the firm's assets falls to a lower threshold value, or called the reorganization boundary (Black and Cox, 1976). Various formulations of default-triggering mechanisms have been developed in the literature. This line of research has been continued with various refinements in the definition of the reorganization boundary, for example, in the recent works by Longstaff and Schwartz (1995), Rich (1996), Briys and de Varenne (1997). Anderson and Tu (1998) proposed the strategic contingent claims models where the reorganization of the firm is also dependent upon actions taken either by the debtors or creditors. These methodologies, though mainly developed for the valuation of risk premia of corporate bonds, can also be adopted to the risk analysis models of swaps.

2.1 Assumptions in the defaultable currency swap models

A currency swap involves exchange of principals and interest payments in two currencies. The counterparties of a currency swap are companies in two different countries; say, let company A be domestic and company B be foreign. We assume company A to be risky and company B to be default free. The firm value F of company A is assumed to follow the stochastic process

$$dF = \mu_F F \; dt + \sigma_F F \; dZ_F \quad (2.1)$$

where μ_F is the instantaneous expected rate of return on the firm's assets, σ_F^2 is the instantaneous variance rate and dZ_F is the standard Wiener process. Further, it is assumed that the value of the firm is independent of the capital structure of the firm. Also, the value of the swap contract is assumed to be negligibly small compared to the firm value F.

Following the intertemporal default model proposed by Longstaff and Schwartz (1995), we assume that there is constant threshold value H for the firm value below which financial distress occurs. That is, the firm value F must be greater than H in order that the firm continues to be able to meet its contractual obligations. The exchange rate S, which is defined as the domestic currency price of one unit of foreign currency, is also assumed to follow the lognormal process

$$dS = \mu_S S \, dt + \sigma_S S \, dZ_S. \quad (2.2)$$

where μ_S and σ_S^2 are the expected rate of return and variance rate of the exchange rate, respectively.

Cashflows between currency swap counterparties

The domestic company A has comparative advantage in borrowing domestic loan but it wants to raise foreign capital, while situations for the foreign company B happen to be reverse to those of company A. It then becomes natural for both companies to enter into a currency swap so as to exploit the comparative advantages in borrowing rates. Let P_d and P_f denote the principals in domestic and foreign currencies, respectively, where $P_d = S_0 P_f$ and S_0 is the exchange rate at the beginning of the swap contract. When the currency swap is initiated, company A pays to company B the principal P_d, in exchange for receiving P_f. During the period of the swap contract, A makes interest payments to B in foreign currency since A has received P_f from B; and vice versa, B makes domestic interest payments to A. For simplicity of analysis, we assume that these payments are in continuous streams at the constant swap rates c_d and c_f, that is, A pays $c_f P_f$ continuously to B while B pays $c_d P_d$ continuously to A throughout the whole life of the swap. The swap rates are chosen such that the value of the swap contract is set to be zero at the beginning of the contract. At maturity of the swap, company A exchanges P_f for P_d with company B. The cashflows between the two currency swap counterparties are summarized in Figure 1.

Settlement rules

When the firm value F of company A falls to the threshold value H, company A is forced to reorganize. The settlement payment to the swap counterparty upon intertemporal default depends on the settlement clauses in the swap contract. Under the full (limited) two-way payment clause, the non-defaulting counterparty is required (not required) to pay if the final net amount is favorable to the defaulting party. When company A defaults and the swap is favorable to its counterparty, company B will receive only the fraction $1 - w$ of the market quotation value of the swap agreement. Here, w denotes the proportion of write-down upon default, and the market quotation value refers to the value of the corresponding riskless swap contract.

2.2 Governing equations

Let $v(S, t)$ and $V(S, F, t)$ denote the values of the riskless and defaultable currency swap contracts to company B, respectively. The governing equations for $v(S, t)$ and $V(S, F, t)$ are derived using the standard riskless hedging argument, and the auxiliary conditions are developed by modelling the cashflow settlements at maturity and upon intertemporal default.

2.2.1 Riskless currency swap models

The governing equation for $v(S,t)$ is given by

$$\frac{\partial v}{\partial t} + \frac{\sigma_S^2}{2} S^2 \frac{\partial^2 v}{\partial S^2} + (r_d - r_f) S \frac{\partial v}{\partial S} + (P_f c_f S - P_d c_d) - r_d v = 0,$$
$$0 < S < \infty, \quad t > 0, \qquad (2.3)$$

where r_d and r_f are the domestic and foreign riskless interest rates, respectively. At maturity, the payoff is

$$v(S,T) = P_f S - P_d. \quad (2.4)$$

For a given domestic swap rate c_d, the foreign swap rate c_f is chosen such that the value of the riskless swap agreement is zero at the initiation of the agreement.

2.2.2 Defaultable currency swap models

The governing equation for $V(S,F,t)$ is given by

$$\frac{\partial V}{\partial t} + \frac{\sigma_S^2}{2} S^2 \frac{\partial^2 V}{\partial S^2} + \rho_{SF} \sigma_S \sigma_F SF \frac{\partial^2 V}{\partial S \partial F} + \frac{\sigma_F^2}{2} F^2 \frac{\partial^2 V}{\partial F^2}$$
$$+ [r_d F - (P_f c_f S - P_d c_d)] \frac{\partial V}{\partial F} + (r_d - r_f) S \frac{\partial V}{\partial S} + (P_f c_f S - P_d c_d) - r_d V = 0,$$
$$0 < S < \infty, H < F < \infty, t > 0, \qquad (2.5)$$

where ρ_{SF} is the correlation coefficient between S and F. The prescription of the auxiliary conditions depends on the settlement clauses of the swap agreement upon default. The limited and full two-way settlement clauses are both considered here.

Limited two-way settlement
(i) At maturity, the two counterparties exchange their principals when A is non-defaulting. When A becomes default, B receives $1 - w$ of $P_f S_T - P_d$ from A if $P_f S_T - P_d > 0$, but B pays nothing to A if $P_f S_T - P_d \leq 0$.

$$V(S,F,T) = \begin{cases} P_f S - P_d, & F > H \\ (1-w)\max(P_f S - P_d, 0), & F = H \end{cases} \cdot \quad (2.6a)$$

(ii) When the firm value F tends to infinity, it is almost sure that F will not fall below H at subsequent times; hence,

$$\lim_{F \to \infty} V(S,F,t) = v(S,t), \quad \text{for all } t. \quad (2.6b)$$

(iii) When the firm value F drops to the threshold value H, company A becomes default. The counterparty B pays nothing if the swap is favorable to the defaulting party A; otherwise, it receives $1 - w$ of the value of the corresponding riskless swap contract. Hence,

$$V(S,H,t) = (1-w)\max(v(S,t),0), \quad \text{for all } t. \quad (2.6c)$$

(iv) When the exchange rate S drops to zero, it will stay at that level at all subsequent times. The foreign payments becomes worthless, and the swap contract behaves like a bond where company B pays the continuous payments $c_d P_d$ and final par value P_d to company A. The present value of these payments equals to $P_d \left\{ e^{-r_d(T-t)} + \frac{c_d}{r_d}[1 - e^{-r_d(T-t)}] \right\}$, provided that company A remains non-defaulting. When company A defaults, company B automatically stops the payments due to the limited two-way settlement clause. Hence,

$$V(0,F,t) = -P_d \left\{ e^{-r_d(T-t)} + \frac{c_d}{r_d}[1 - e^{-r_d(T-t)}] \right\} P[F \geq H], \quad (2.6d)$$

where $P[F \geq H]$ denotes the probability that the firm value F stays above H for the whole life of the swap contract. It is known that

$$P[F \geq H] = N\left(\frac{\ln\frac{V}{H} + \left(r_d - \frac{\sigma_V^2}{2}\right)(T-t)}{\sigma_F\sqrt{T-t}}\right)$$
$$- e^{-\frac{2r_d}{\sigma_V^2}\ln\frac{V}{H}} N\left(\frac{\ln\frac{H}{V} + \left(r_d - \frac{\sigma_V^2}{2}\right)(T-t)}{\sigma_F\sqrt{T-t}}\right). \quad (2.6e)$$

(v) It is quite tricky to prescribe the far field boundary condition at $S \to \infty$. Instead of adopting an artificial boundary condition, we use the skew discretization technique where the discretization of the governing equation along the boundary nodes involves lattice nodes which are completely inside the computational domain.

Full two-way settlement
Under the full two-way settlement clause, B has to honor the swap contract even when A becomes default. The auxiliary conditions which model the payments paid by B upon the default of A have to be modified as follows:
(i) At maturity,

$$V(S, F, T) = \begin{cases} P_fS - P_d & F>H \\ P_fS - P_d & F = H \text{ and } P_fS - P_d \leq 0 \\ (1-w)(P_fS - P_d) & F=H \text{ and } P_fS - P_d > 0 \end{cases}. \quad (2.7a)$$

(ii) When the firm value F hits H,

$$V(S, H, t) = \begin{cases} (1-w)v(S,t) & v(S, t)>0 \\ v(S,t) & v(S, t) \leq 0 \end{cases}. \quad (2.7b)$$

(iii) When the exchange rate S drops to zero,

$$V(0, F, t) = -P_d\left\{e^{-r_d(T-t)} + \frac{c_d}{r_d}[1 - e^{-r_d(T-t)}]\right\}. \quad (2.7c)$$

Once the governing equations for the currency swap models are well formulated with full prescription of the auxiliary conditions, the equations can be solved numerically by standard finite difference method.

3. CHARACTERIZATION OF SWAP RATE SPREADS

We would like to analyze the impact on the swap rate spread under the influence of (i) firm credit rating, (ii) volatility of the firm value (iii) volatility of the exchange rate, (iv) correlation between firm value and exchange rate. Here, the swap spread is defined as the difference of the foreign swap rates c_f with and without the default possibilities of party A (recall that party B is assumed to be default free). Also, we examine the effects of the limited and full two-way settlement clauses on the swap spreads.

Recall that the foreign swap rate c_f gives the rate of continuous foreign payments from A to B. Suppose a settlement clause is more (less) favorable to party B, we would expect a decrease (increase) in c_f. Equivalently, this leads to negative (positive) swap spread on c_f in order that the swap contract is fair to both parties.

Firstly, we investigate the relationship between the swap spread on c_f and the credit rating of the defaultable party. Figure 2 shows the plots of swap spreads on c_f against F/H. For the given set of parameter values chosen in the swap model, the swap contract is in-the-money to the defaultable party A. It is observed that the swap spreads are negative for the limited two-way payment clause. This is expected as the limited settlement clause gives advantage to party B since B can be excused from honoring an out-of-the-money swap contract to itself when A defaults. The swap spread narrows as the credit rating of A improves, and becomes essentially zero as the ratio F/H goes beyond 3. For the full two-way settlement, the swap spread curve reveals that the swap spreads are always positive and the spread achieves a maximum value at certain level of F/H. The positivity of the swap spreads reflects the replacement cost to the non-defaulting party since it receives only a fraction of the market quotation of the swap contract upon default of the counterparty. Since the swap contract is in-the-money to the defaulting party A, the loss to counterparty B is zero when A defaults. Hence, the swap spread value becomes zero when F hits H. This explains why the swap spread curve corresponding to full two-way settlement increases from zero to some maximum value then decreases asymptotically to zero value at high firm value.

Secondly, we examine the relationship between the swap spread on c_f and the volatility of the firm value of A, σ_F (see Figure 3). When the firm value becomes more volatile, the possibility of default increases and the swap spreads become widened, that is, more negative for the limited two-way payment and more positive for the full two-way payment. The swap spreads tend to some asymptotic values at high σ_F.

Thirdly, the relationship between the swap spread and the volatility of the exchange rate is revealed in Figure 4. With the correlation coefficient between the firm value and exchange rate being positive, the firm has a higher possibility to default when the exchange rate becomes more volatile. Therefore, the value of swap spread increases in magnitude at higher value of σ_F for both full and limited two-way settlement clauses.

Lastly, the correlation coefficient ρ_{SF} is shown to have significant impact on the swap spreads (see Figure 5). For both full and limited two-way settlement clauses, the swap spreads are decreasing functions of the correlation coefficient. For full two-way settlement, the swap spread is always positive since B always loses upon the default of the counterparty A. Suppose the correlation is highly positive and S increases, there is a higher tendency for the increase of F (that is, less susceptible to default) so the expected loss to B associated with the replacement cost drops. On the other hand, when S decreases, the swap becomes more in-the-money to A. Correspondingly, the expected replacement cost incurs to B when A becomes default tends to a small value. Both arguments explain why the swap spread decreases when the correlation coefficient increases. For limited two-way settlement, the above argument of drop in swap spread with more positive correlation still applies, except that the swap spread can decrease beyond the zero value at high positive correlation.

4. CONCLUSIONS

In this paper, the contingent claim approach is employed to analyze the impact on the swap spreads of a currency swap with default risk of one of the counterparties. This paper goes beyond previous similar works by allowing intertemporal default of the defaultable party. It is observed that the limited and full two-way settlement clauses have significant effects on the behaviors of the swap spreads. The swap spreads depend sensibly on the proximity of the firm value to the defaulting threshold value, the volatilities of the firm value and exchange rate, and the correlation coefficient between firm value and exchange rate. It is observed that the full two-way settlement clause is always unfavorable to the non-defaulting party B so this leads to positive swap spread of the foreign swap rate (that is, defaultable party A should pay more in continuous foreign payments as compensation). The situation becomes more complicated for the limited two-way settlement clause. The non-defaulting party B may face losses (replacement cost) upon the default of the counterparty A. On the other hand, B is not required to honor the swap payment to A upon the default of A. Hence, the swap spread corresponding to the limited two-way settlement clause can be either positive or negative, depending on the relative strengths of the above two competing factors.

References

[1] Anderson, R.W. and C. Tu, "Numerical analysis of strategic contingent claims models," *Computational Economics*, vol. 11 (1998) p.3-19.

[2] Baz, J. and M.J. Pascutti, "Alternative swap contracts: analysis and pricing," *Journal of Derivatives*, winter issue (1996) p.7-21.

[3] Briys, E. and F. de Varenne, "Valuing risky fixed rate debt: an extension" *Journal of Financial and Quantitative Analysis*, vol. 32 (1997) p.239-248.

[4] Black, F. and J.C. Cox, "Valuing corporate securities: some effects of bond indenture provisions," *Journal of Finance*, vol. 31 (1976) p.351-367.

[5] Cooper, I.A. and A.S. Mello, "The default risk of swap," *Journal of Finance*, vol. 46 (1991) p.597-620.

[6] Jarrow, R. and S. Turnbull, "When swaps are dropped," *Risk*, vol. 10 (May, 1997) p.70-74.

[7] Li, H.,"Pricing of swaps with default risk," *Review of Derivatives Research*, vol. 2 (1998) p.231-250.

[8] Litzenberger, R.H., "Swaps: plain and fanciful," *Journal of Finance*, vol. 47 (1992) p.831-850.

[9] Longstaff, F.A. and E.S. Schwartz, "A simple approach to valuing risky fixed and floating rate debt," *Journal of Finance*, vol. 50 (1995) p.789-819.

[10] Merton, R.C., "On the pricing of corporate debt: The risk structure of interest rates," *Journal of Finance*, vol. 29 (1974) p.449-470.

[11] Nielsen, S.S. and E.I. Ronn, "The valuation of default risk in corporate bonds and interest rate swaps," *Advances in Futures and Options Research*, vol. 9 (1997) p.175-196.

[12] Rich, D. and R. Leipus, "An option-based approach to analyzing financial contracts with multiple indenture provisions," *Advances in Futures and Options Research*, vol. 9 (1997) p.1-36.

DEPARTMENT OF INFORMATION SYSTEMS,SCHOOL OF COMPUTING,NATIONAL UNIVERSITY OF SINGAPORE, SINGAPORE 119260

DEPARTMENT OF MATHEMATICS, HONG KONG UNIVERSITY OF SCIENCE AND TECHNOLOGY, HONG KONG

FIGURES

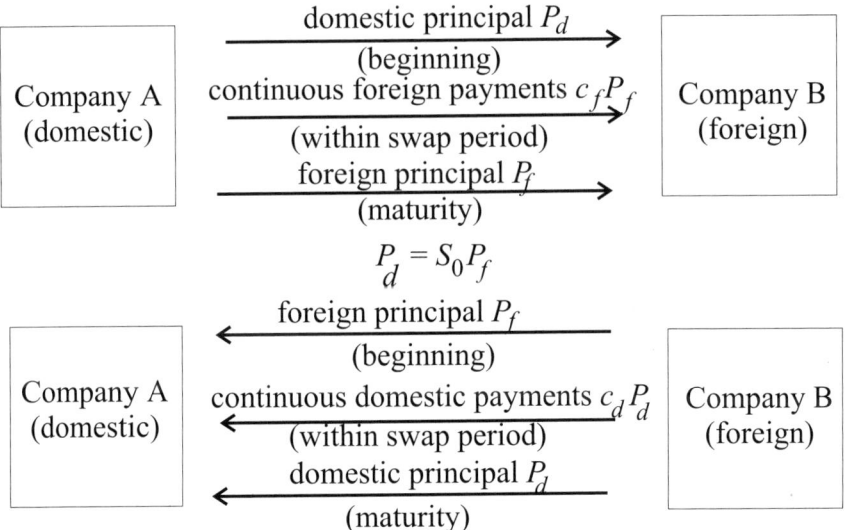

FIGURE 1. Cashflows between the two currency swap counterparties, assuming no intertemporal default.

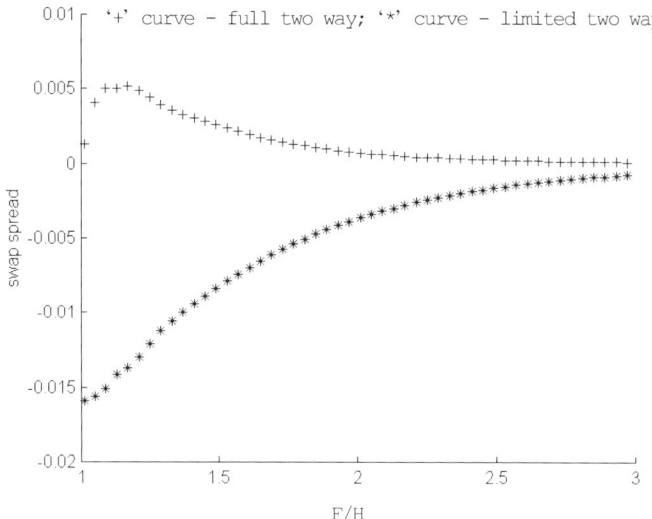

FIGURE 2. The relationship between the swap spread on c_f and the firm credit rating as measured by F/H, where F is the firm value and H is the default level. The parameter values are $P_d = 1, S_0 = 2, S = 2, c_d = 8\%, T = 4, r_d = r_f = 6\%, \sigma_S = 15\%, \sigma_F = 25\%, w = 0.75$ and $\rho_{SF} = 0.25$.

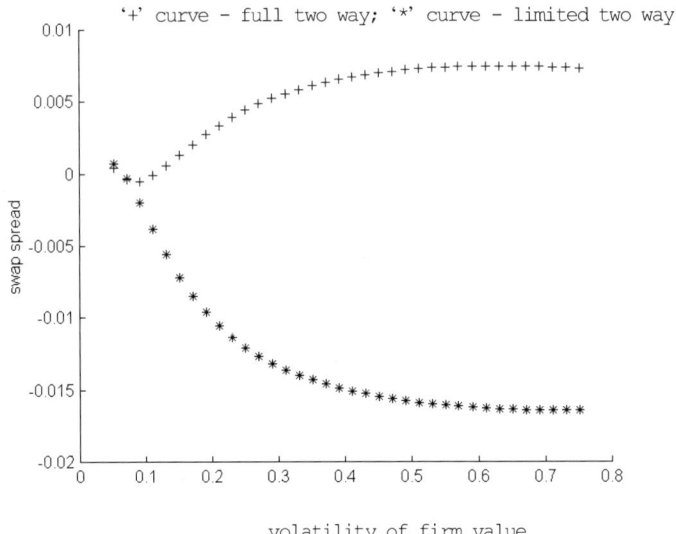

FIGURE 3. The relationship between the swap spread on c_f and the volatility of the firm value, σ_F. The same set of parameter values are used as in Figure 2, except that σ_F is allowed to vary and $H = 100, F = 125$.

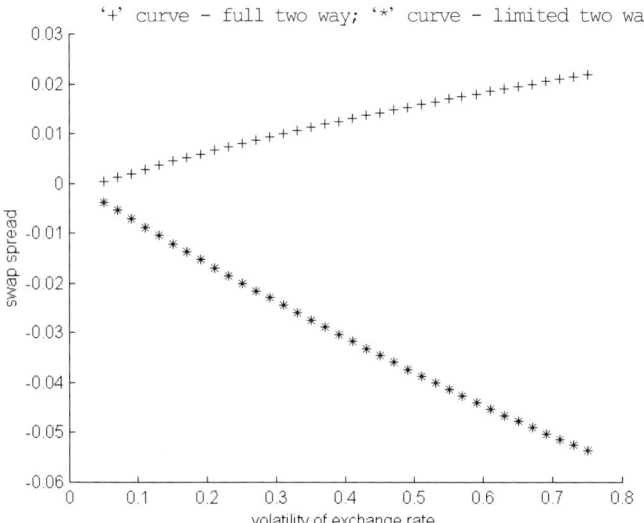

FIGURE 4. The relationship between the swap spread on c_f and the volatility of the exchange rate, σ_S. The same set of parameter values are used as in Figure 3, except that $\sigma_F = 25\%$ and σ_S is allowed to vary.

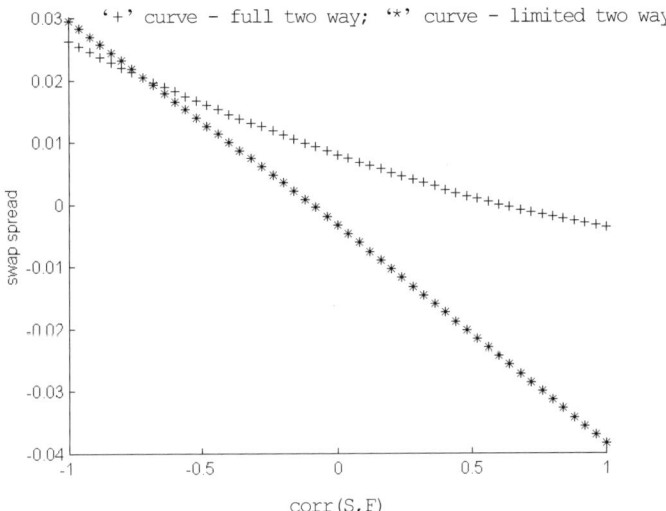

FIGURE 5. The relationship between the swap spread on c_f and the correlation coefficient between exchange rate and firm value, ρ_{SF}. The same set of parameter values are used as in Figures 3 and 4, except that ρ_{SF} is allowed to vary.

AMS/IP Studies in Advanced Mathematics
Volume 26, 2002

Dynamic Insider Trading

Shunlong Luo and Qiang Zhang

ABSTRACT. We generalize Kyle's paradigm of dynamic insider trading to a more information relevant setting. A single risky asset is traded among a risk neutral insider and noise traders with the intermediation of competitive risk neutral market makers. The insider acquires some private information about the future value of the risky asset, and takes a speculative position based on this information. The noise traders submit a random order. The market makers set the price conditional on the total order in a rational expectations fashion. The linear equilibrium is characterized analytically and the impact of the inside information on the market is quantified by virtue of stochastic filtering theory.

1. Introduction

In the study of financial markets, homogeneous (symmetric) information are usually implicitly or explicitly assumed, this is not realistic since heterogeneous and asymmetric information among individuals are the rule rather than the exception, they constitute an important aspect of financial markets. Considerable evidences support this claim, e.g., the existence of active markets for information, such as advisory services and newsletters, the regulation of insider trading. Many mutual fund managers claim to trade on the basis of superior information. Numerous markets are characterized by informational difference between buyers and sellers. In banking market, borrowers usually possess more private information such as their own characteristics than lenders. In principal-agent relationship, the entrepreneurs possess "inside information" about their own projects which they seek financing, while adverse selection hampers the efficient transfer of information. Market psychology and consumer sentiment, which have little to do with the fundamentals of economy, also contribute to heterogeneity of beliefs which leads to the information asymmetries.

In the last twenty years, considerable efforts are devoted to modeling and quantifying information impact on financial markets. Developments of informational

Key words and phrases. Information; Insider; Noise traders; Semi-strong efficiency; Nash equilibrium.

Supported by City University of Hong Kong, contracts 7000776, 8360020, 9000906 and 9030641, and RGC contract 9040399.

economics pertaining to the theory of finance include the role of prices in aggregating and communicating information, the signaling of asymmetrically known attributes, the revelation of information through contingent contracting, and the analysis of strategic coordination across multiple informed agents.

Among the above, one of the most active areas is the information aggregation and its incorporation into prices in rational expectations markets. Grossman and Stiglitz (1980) studied a lot of quantitative properties about the rational expectations equilibrium when the price conveys information in a Gaussian and constant absolute risk utility framework. Hellwig (1980) developed a large-market model to study the information aggregation mechanism. Diamond and Verrecchia (1981) studied the information aggregation in a noisy rational expectations economy. Verrecchia (1982) presented a model of information acquisition in the context of Hellwig's large-market competitive model. He showed that there exists a rational expectations equilibrium with endogenously determined information cost. Admati (1985) extended Hellwig's model to a multi-asset situation and discovered many new phenomena due to the interaction among different assets. Wang (1993) established an intertemporal equilibrium model of asset pricing under asymmetric information. Brennan and Cao (1996), worked in the Hellwig's noisy rational economy, analyzed the value of improving trading opportunities by more frequent trading in which the information can be more thoroughly exploited.

In his insightful and influential paper, Kyle (1985) established a model of insider trading and showed how a *perfectly* informed insider can make speculative profit from his private information by trading strategically and by hiding his private information in the camouflage of noise traders. Since then, this model has become a workhorse for analyzing strategic insider trading and market microstructure. Many authors have used variants of this model to analyze and to explain real financial phenomena. Foe example, Kyle (1989), Campbell and Kyle (1993) further developed some related models to study informed speculation with imperfect information, and explored the interaction of exogenous noise traders with "smart money" investors who have constant absolute risk aversion. Back (1992) presented an elegant continuous version of Kyle's model with a single insider in the framework of a Brownian martingale. Admati and Pfleiderer (1988) examined some interday trading patterns by means of Kyle's model. Spiegel and Subrahmanyam (1992) replaced the price inelastic noise traders (liquidity traders) with strategic, utility-maximizing hedgers in Kyle's static model and endogenized the trading motives of these agents. Holden and Subrahmanyam (1992) used Kyle's multi-period model to study the imperfect competition among insiders. See O'Hara (1995) for a survey of many applications of Kyle's model in financial markets.

In this paper, we extend Kyle's dynamic model to incorporate more realistic inside information and study its implications. In contrast to Kyle's model in which the insider knows *exactly* the future value of the asset, we allow the information to be partial, i.e., the insider may only obtain a *noisy* version of the future value. We use stochastic filtering theory to clarify the derivation of the dynamic continuous time trading equilibrium. This *complements and generalizes* Kyle's model. The principal results includes explicit characterizations of the linear equilibrium and the various information relevant quantities.

The paper is structured as follows. In Section 2, we describe the dynamic insider trading model, and prepare a stochastic filtering lemma. In Section 3, we derive the unique linear Nash equilibrium and explores its informational implications. Section 4 concludes.

2. The Dynamic Insider Trading

All uncertainty is supported on a standard probability space $(\Omega, \mathcal{F}, \mathrm{P})$, with E, cov and var denoting expectation, covariance and variance, respectively. A piece of information in our context is a signal, a realization of a random variable that is correlated with the state of the world, more specifically, correlated with the risky asset future value. The precision of the information is parameterized by the variance.

Now we describe the financial market. The trading model is a natural generalization of Kyle's (1985) to imperfect private information, it consists of the following ingredients.

Trading horizon: The trading takes place continuously over the time interval $[0, 1)$. At time 1, the uncertainty is resolved and the risky asset payoff is realized. The risk free interest rate is normalized to zero.

Asset: There is a single risky asset whose *ex post* liquidate value at time 1 is a random variable $\widetilde{v}$, normally distributed with mean 0 and variance Σ_0. Here the assumption of a zero mean is for simplicity, a non-zero mean can be treated similarly.

Traders: The economy has three kinds of traders: a single *risk neutral* insider, noise traders, and competitive *risk neutral* market makers. The insider has superior private information about the liquidate value of the risky asset, and rationally anticipates the effect of his trading strategies on the prices. Noise traders (liquidity traders) purchase a random, exogenous, inelastic quantity of the risky asset. Their motives for trading are irrelevant for our analysis here. But their presence is crucial for the model since this serves as a camouflage and makes it impossible for the market makers to exactly infer the insider's information. The noise traders are also the source of profits to be exploited by insiders. The insider and noise traders are assumed to submit their orders to market makers who set prices efficiently in a rational expectations fashion. The market makers expect to earn a zero profit.

The trading proceeds as follows: At time 0, the insider observes a signal

$$\widetilde{s} = \widetilde{v} + \widetilde{\epsilon},$$

where the noise variable $\widetilde{\epsilon}$ is a normal random variable with mean 0 and variance σ (we use σ rather than σ^2 to denote variance for notational simplicity.) In addition, $\widetilde{\epsilon}$ is assumed to be independent of $\widetilde{v}$. This form of signal is the standard "signal equals fundamental plus noise", which is widely used in literature. The inverse of σ reflects the precision of the information. At every time t, the insider chooses his trading strategy by submitting an order $dX(t)$ (rate of trading) based on his observed signal $\widetilde{s}$ and the price history. Thus the cumulative order process of the insider is $\{X(t)\}_{0 \le t < 1}$. Let $\{\sqrt{\Sigma}\widetilde{u}(t)\}_{0 \le t < 1}$ be the cumulative order process of the noise traders. Here $\{\widetilde{u}(t)\}_{0 \le t < 1}$ is a standard Brownian motion (independent of $\widetilde{v}$ and $\widetilde{\epsilon}$), and Σ is a scale parameter. $\sqrt{\Sigma}\widetilde{u}(t)$ represents the cumulative order of the noise traders up to time t. Thus the cumulative aggregate order of all insiders and noise traders up to time t is $X(t) + \sqrt{\Sigma}\widetilde{u}(t)$. After receiving the aggregate order, the market makers determine the price $P(t)$ of the risky asset at which they trade the quantity necessary to clear the market. When doing so they observe the aggregate order $X(t) + \sqrt{\Sigma}\widetilde{u}(t)$, but cannot disentangle the amount, that is, they cannot observe $X(t)$ or $\sqrt{\Sigma}\widetilde{u}(t)$ separately (nor can they observe $\widetilde{s}$ or $\widetilde{v}$). Consequently, the price is not fully revealing. The trading takes place continuously until time 1 (not inclusive).

We do not pursue the mechanism by which the market makers determining the price. It would be helpful to imagine that the efficient price will result from a Bertrand competition of many risk neutral market makers, in which their profits are driven to zero. The assumption of normality of random variables is for simplicity and analytical tractability, as is standard in the literature.

Let $\{\mathcal{F}(t)\}_{0 \leq t < 1}$ be the natural filtration generated by the aggregate order process $\{X(t) + \sqrt{\Sigma}\widetilde{u}(t)\}_{0 \leq t < 1}$ (thus, $\mathcal{F}(t)$ is the information available to the market makers up to time t, and $\mathcal{F}(0)$ is trivial). In the conjectured equilibrium, this filtration is the same as that generated by the price process $\{P(t)\}_{0 \leq t < 1}$ as will be seen from Eq. (2) in the sequel. The insider's *ex ante* expected profit over the whole time period is

$$\Pi(X, P) \equiv \int_0^1 \left(\widetilde{v} - P(t)\right) dX(t).$$

Our object is to find the optimal trading strategy of the (speculative) insider and the equilibrium pricing rule of the market makers, and to explore the impact of the inside information on the market.

A *(Nash) equilibrium* in the economy is a trading strategy $\{X(t)\}_{0 \leq t < 1}$ for the insider, and a pricing rule $\{P(t)\}_{0 \leq t < 1}$ for the market makers such that

(1) for the insider's alternate trading strategy $\{X'(t)\}_{0 \leq t < 1}$

$$\Pi(X, P) \geq \Pi(X', P);$$

(2) the market is semi-strong efficient

$$\mathrm{E}[\widetilde{v} | \mathcal{F}(t)] = P(t), \qquad t \in [0, 1).$$

REMARK. Following the original formulation of Kyle (1985), it seems more reasonable to replace condition (1) by

$$\mathrm{E}\left[\int_\tau^1 (\widetilde{v} - P(t)) dX(t) \Big| \{P(t)\}_{0 \leq t \leq \tau}, \widetilde{s}\right] > \mathrm{E}\left[\int_\tau^1 (\widetilde{v} - P(t)) dX'(t) \Big| \{P(t)\}_{0 \leq t \leq \tau}, \widetilde{s}\right]$$

for any $\tau \in [0, 1)$. But as remarked by Kyle (1985), since we only consider linear strategies and linear pricing rules, we need only consider (1).

The model described above is a natural extension of Kyle's model (1985). When $\sigma \to 0$, we recover Kyle's model (1985) with continuous time.

We will use the following filtering theorem taken from Lipster and Shirayaev (1977).

LEMMA 1. *Let $W_1(t)$ be a k-dimension standard Brownian motion and $W_2(t)$ an l-dimension standard Brownian motion. Let $\theta_t = (\theta_1(t), \cdots, \theta_k(t))^*$ be a k-dimension process and $\xi(t) = (\xi_1(t), \cdots, \xi_l(t))^*$ an l-dimension process. Suppose that they satisfy the following stochastic differential equations*

$$d\theta(t) = [a_0(t) + a_1(t)\theta(t) + a_2(t)\xi(t)]dt + b_1(t)dW_1(t) + b_2(t)dW_2(t),$$
$$d\xi(t) = [A_0(t) + A_1(t)\theta(t) + A_2(t)\xi(t)]dt + B_1(t)dW_1(t) + B_2(t)dW_2(t).$$

Here $a_0(t)$, $a_1(t)$, $a_2(t)$, $b_1(t)$ and $b_2(t)$ are $k \times 1$, $k \times k$, $k \times l$, $k \times k$ and $k \times l$ matrices, respectively; $A_0(t), A_1(t), A_2(t), B_1(t)$ and $B_2(t)$ are $l \times 1, l \times k, l \times l, l \times k$

and $l \times l$ matrices respectively. Let $\mathcal{F}(t)$ be the filtration generated by the process $\xi(t)$, and

$$
\begin{aligned}
m(t) &= \mathrm{E}[\theta(t)|\mathcal{F}(t)] = (\mathrm{E}[\theta_1(t)|\mathcal{F}(t)], \cdots, \mathrm{E}[\theta_k(t)|\mathcal{F}(t)])^*, \\
\gamma(t) &= \mathrm{E}[(\theta(t) - \mathrm{E}[\theta(t)|\mathcal{F}(t)])^*(\theta(t) - \mathrm{E}[\theta(t)|\mathcal{F}(t)])].
\end{aligned}
$$

Then

$$
\begin{aligned}
dm(t) &= a_0(t) + a_1(t)m(t) + a_2(t)\xi(t)]dt + [(b \circ B)(t) + \gamma(t)A_1^*(t)](B \circ B)^{-1}(t) \\
&\qquad \times [d\xi(t) - (A_0(t) + A_1(t)m(t) + A_2(t)\xi(t))dt], \\
\frac{d\gamma(t)}{dt} &= a_1(t)\gamma(t) + \gamma(t)a_1^*(t) + (b \circ b)(t) - [(b \circ B)(t) + \gamma(t)A_1^*(t)](B \circ B)^{-1}(t) \\
&\qquad \times [(b \circ B)(t) + \gamma(t)A_1^*(t)]^*.
\end{aligned}
$$

Here

$$
\begin{aligned}
(b \circ b)(t) &= b_1(t)b_1^*(t) + b_2(t)b_2^*(t), \\
(b \circ B)(t) &= b_1(t)B_1^*(t) + b_2(t)B_2^*(t), \\
(B \circ B)(t) &= B_1(t)B_1^*(t) + B_2(t)B_2^*(t).
\end{aligned}
$$

3. The Equilibrium

Following Kyle (1985), for analytical tractability, we only consider equilibria in which both the trading strategies of the insider and the pricing rules of the market makers are linear functionals. It is an open problem about the situation of nonlinear equilibria. However, Rochet and Vila (1994) proved the uniqueness of equilibrium in a different static model in which the insider observes the amount of the noise traders' order (Kyle's original model does not possess this property).

We first conjecture a linear (Nash) equilibrium and then verify that it is indeed an equilibrium. Inspired by Kyle (1985), suppose that the trading strategy of the insider and the pricing rule of the market makers are described by the following stochastic differential equations:

$$
\begin{aligned}
dX(t) &= [\alpha(t)\widetilde{s} - \beta(t)P(t)]dt, & (1) \\
dP(t) &= \lambda(t)[dX(t) + \sqrt{\Sigma}d\widetilde{u}(t)], & (2)
\end{aligned}
$$

respectively. Here $\alpha(t)$, $\beta(t)$, $\lambda(t)$ are deterministic *positive* functions.

Note that there is an important difference between Eq. (1) here and Eq. (4.2) in Kyle (1985). In Kyle (1985), $\alpha(t) = \beta(t)$, but here $\alpha(t) \neq \beta(t)$. The difference is due to that in Kyle (1985), the inside information is perfect ($\sigma = 0$), but here the inside information may not be perfect, that is, σ may not be zero.

Before we explicitly derive the Nash equilibrium, we justify the semi-strong efficiency by showing that the market makers will indeed make a zero profit in equilibrium.

THEOREM 2. *In equilibrium, the market makers make a zero profit, that is*

$$
\mathrm{E} \int_0^1 (\widetilde{v} - P(t))d(-X(t) - \sqrt{\Sigma}\widetilde{u}(t)) = 0.
$$

Proof. By the condition of market semi-strong efficiency, $\{P(t)\}_{0 \leq t < 1}$ is a martingale under the filtration $\{\mathcal{F}(t)\}_{0 \leq t < 1}$ and $P(0) = \mathrm{E}[\widetilde{v}|\mathcal{F}(0)] = \mathrm{E}[\widetilde{v}] = 0$.

Since by Eq. (1)

$$d(X(t) + \sqrt{\Sigma}\widetilde{u}(t)) = \frac{1}{\lambda(t)}dP(t),$$

the process

$$\left\{ Z_\tau \equiv \int_0^\tau (\widetilde{v} - P(t))\frac{1}{\lambda(t)}dP(t) \right\}_{0 \le \tau < 1}$$

is an $\{\mathcal{F}(t)\}_{0 \le t < 1}$−martingale and $X_0 = 0$. Consequently,

$$\mathrm{E}\int_0^1 (\widetilde{v} - P(t))d(-X(t) - \sqrt{\Sigma}\widetilde{u}(t))$$

$$= -\mathrm{E}\int_0^1 (\widetilde{v} - P(t))\frac{1}{\lambda(t)}dP(t)$$

$$= -\mathrm{E}[Z_1] = -\mathrm{E}[Z_0] = 0.$$

The following is our main result.

THEOREM 3. *The unique linear equilibrium in the continuous time trading is determined by Eqs. (1) and (2) with*

$$\alpha(t) = \sqrt{\frac{\Sigma}{\Sigma_0 + \sigma}}\frac{1}{1-t}$$

$$\beta(t) = \frac{\sqrt{\Sigma(\Sigma_0 + \sigma)}}{\Sigma_0}\frac{1}{1-t}$$

$$\lambda(t) = \frac{\Sigma_0}{\sqrt{\Sigma(\Sigma_0 + \sigma)}}.$$

That is, the equilibrium trading strategy of the insider and the equilibrium pricing rule of the market makers are determined by

$$dX(t) = \left(\sqrt{\frac{\Sigma}{\Sigma_0 + \sigma}}\frac{1}{1-t}\widetilde{s} - \frac{\sqrt{\Sigma(\Sigma_0 + \sigma)}}{\Sigma_0}\frac{1}{1-t}P(t) \right)dt,$$

$$dP(t) = \frac{\Sigma_0}{\sqrt{\Sigma(\Sigma_0 + \sigma)}}\left(dX(t) + \sqrt{\Sigma}d\widetilde{u}(t) \right),$$

respectively.

Proof. The method to derive the equilibrium is simple: First, by virtue of (filtering) Lemma 1, we compute $\mathrm{E}[\widetilde{v}|\mathcal{F}(t)]$. By market semi-strong efficiency, comparing $\mathrm{E}[\widetilde{v}|\mathcal{F}(t)]$ with $P(t)$ postulated by Eq. (2), we obtain two equations relating $\alpha(t)$, $\beta(t)$ and $\lambda(t)$ (note that the filtering theory is used by Wang (1993) in another content to derive a rational expectations equilibrium). Second, we solve the profit maximization problem for the insider, and come to an additional equation. From these two steps we can determine the equilibrium parameter functions. The detail is as follows.

Since $\widetilde{s} = \widetilde{v} + \widetilde{\epsilon}$, the price differential equation (2) can be rewritten as

$$dP(t) = \lambda(t)\left(\alpha(t)(\widetilde{v} + \widetilde{\epsilon})dt - \beta(t)P(t) + \sqrt{\Sigma}d\widetilde{u}(t) \right)$$

$$= \lambda(t)\alpha(t)(1,1)\begin{pmatrix} \widetilde{v} \\ \widetilde{\epsilon} \end{pmatrix}dt - \lambda(t)\beta(t)P(t)dt + \sqrt{\Sigma}\lambda(t)d\widetilde{u}(t).$$

We first consider the filtering of $\widetilde{v}$ and $\widetilde{\epsilon}$ with respect to $\{\mathcal{F}(t)\}_{0 \leq t < 1}$. Let

$$\mathbf{V}(t) = \left(\begin{array}{c} \mathrm{E}[\widetilde{v}|\mathcal{F}(t)] \\ \mathrm{E}[\widetilde{\epsilon}|\mathcal{F}(t)] \end{array} \right) = \mathrm{E}\left[\left(\begin{array}{c} \widetilde{v} \\ \widetilde{\epsilon} \end{array} \right) \Big| \mathcal{F}(t) \right].$$

By Lemma 1, we have

$$dV(t)$$
$$= \Gamma(t)\lambda(t)\alpha(t) \left(\begin{array}{c} 1 \\ 1 \end{array} \right) \frac{1}{\lambda^2(t)\Sigma} \left\{ \lambda(t)\alpha(t)(1,1) \left[\left(\begin{array}{c} \widetilde{v} \\ \widetilde{\epsilon} \end{array} \right) - \mathbf{V}(t) \right] dt + \sqrt{\Sigma}\lambda(t)d\widetilde{u}(t) \right\}$$
$$= \frac{1}{\Sigma}\alpha(t)\Gamma(t) \left(\begin{array}{c} 1 \\ 1 \end{array} \right) \frac{1}{\lambda^2(t)\Sigma} \left\{ \alpha(t)(1,1) \left[\left(\begin{array}{c} \widetilde{v} \\ \widetilde{\epsilon} \end{array} \right) - \mathbf{V}(t) \right] dt + \sqrt{\Sigma}d\widetilde{u}(t) \right\}.$$

Here

$$\Gamma(t) = \mathrm{E}\left[\left(\left(\begin{array}{c} \widetilde{v} \\ \widetilde{\epsilon} \end{array} \right) - \mathbf{V}(t) \right) ((\widetilde{v}, \widetilde{\epsilon}) - \mathbf{V}^*(t)) \right]$$

satisfies the following matrix Riccati differential equation

$$\frac{d\Gamma(t)}{dt} = -\Gamma(t)\lambda(t)\alpha(t) \left(\begin{array}{c} 1 \\ 1 \end{array} \right) \frac{1}{\lambda^2(t)\Sigma}\lambda(t)\alpha(t)(1,1)\Gamma(t)$$
$$= -\Gamma(t)\frac{\alpha^2(t)}{\Sigma} \left(\begin{array}{cc} 1 & 1 \\ 1 & 1 \end{array} \right) \Gamma(t),$$
$$\Gamma(0) = \left(\begin{array}{cc} \Sigma_0 & 0 \\ 0 & \sigma \end{array} \right).$$

The solution to the above equation is

$$\Gamma(t) = \left[\Gamma^{-1}(0) + \int_0^t \frac{\alpha^2(t)}{\Sigma}dt \left(\begin{array}{cc} 1 & 1 \\ 1 & 1 \end{array} \right) \right]^{-1}$$
$$= \frac{\Sigma_0\sigma}{1 + (\Sigma_0 + \sigma)a} \left(\begin{array}{cc} \sigma^{-1} + a & -a \\ -a & \Sigma_0^{-1} + a \end{array} \right).$$

Here $a = \int_0^t \frac{\alpha^2(t)}{\Sigma}dt$.

Comparing the two components of the vector process $V(t)$, we have

$$\Sigma_0^{-1}d\mathrm{E}[\widetilde{v}|\mathcal{F}(t)] = \sigma^{-1}d\mathrm{E}[\widetilde{\epsilon}|\mathcal{F}(t)]. \tag{3}$$

Now consider the filtering of $\widetilde{s}$ with respect to $\{\mathcal{F}(t)\}_{0 \leq t < 1}$. By Lemma 1, we have

$$d\mathrm{E}[\widetilde{s}|\mathcal{F}(t)] = \frac{\gamma(t)\lambda(t)\alpha(t)}{\lambda^2(t)\Sigma} \left\{ \lambda(t)\alpha(t)(\widetilde{s} - \mathrm{E}[\widetilde{s}|\mathcal{F}(t)])dt + \lambda(t)\sqrt{\Sigma}d\widetilde{u}(t) \right\}$$
$$= \frac{1}{\Sigma}\alpha(t)\gamma(t) \left\{ \alpha(t)(\widetilde{s} - \mathrm{E}[\widetilde{s}|\mathcal{F}(t)])dt + \lambda(t)\sqrt{\Sigma}d\widetilde{u}(t) \right\}. \tag{4}$$

Here $\gamma(t) = \mathrm{E}[\widetilde{s} - \mathrm{E}[\widetilde{s}|\mathcal{F}(t)]]^2$ satisfies the Riccati differential equation

$$\frac{d\gamma(t)}{dt} = -\gamma(t)\lambda(t)\alpha(t)\frac{1}{\lambda(t)^2\Sigma}\lambda(t)\alpha(t)\gamma(t)$$
$$= -\frac{\alpha^2(t)}{\Sigma}\gamma^2(t),$$
$$\gamma(0) = \Sigma_0 + \sigma.$$

The solution to the above equation is

$$\gamma(t) = \left(\frac{1}{\Sigma_0 + \sigma} + \int_0^t \frac{\alpha^2(t)}{\Sigma} dt\right)^{-1}. \tag{5}$$

On the other hand, since $\widetilde{s} = \widetilde{v} + \widetilde{\epsilon}$, we have

$$d\mathrm{E}[\widetilde{s}|\mathcal{F}(t)] = d\mathrm{E}[\widetilde{v}|\mathcal{F}(t)] + d\mathrm{E}[\widetilde{\epsilon}|\mathcal{F}(t)].$$

Combining this with Eqs. (3) and (4), we have

$$\left(1 + \frac{\sigma}{\Sigma_0}\right)d\mathrm{E}[\widetilde{v}|\mathcal{F}(t)] = \gamma(t)\alpha(t)\frac{1}{\Sigma}\left\{\alpha(t)\left[\widetilde{s} - \left(1 + \frac{\sigma}{\Sigma_0}\right)\mathrm{E}[\widetilde{v}|\mathcal{F}(t)]\right]dt + \sqrt{\Sigma}d\widetilde{u}(t)\right\}.$$

In equilibrium, the semi-strong efficient condition is $\mathrm{E}[\widetilde{v}|\mathcal{F}(t)] = P(t)$. Thus

$$dP(t) = \left(1 + \frac{\sigma}{\Sigma_0}\right)^{-1}\gamma(t)\alpha(t)\frac{1}{\Sigma}\left\{\alpha(t)\left[\widetilde{s} - \left(1 + \frac{\sigma}{\Sigma_0}\right)P(t)\right]dt + \sqrt{\Sigma}d\widetilde{u}(t)\right\}.$$

Comparing this with Eqs. (1) and (2), equating the coefficients of $\widetilde{s}$, $P(t)$ and $d\widetilde{u}(t)$ respectively, and after some manipulation, we obtain

$$\alpha(t) = \left(1 + \frac{\sigma}{\Sigma_0}\right)\Sigma\frac{\lambda(t)}{\gamma(t)}, \tag{6}$$

$$\beta(t) = \left(1 + \frac{\sigma}{\Sigma_0}\right)\alpha(t). \tag{7}$$

From Eqs. (5) and (6), we have

$$\alpha(t) = \left(1 + \frac{\sigma}{\Sigma_0}\right)\Sigma\lambda(t)\left(\frac{1}{\Sigma_0 + \sigma} + \int_0^t \frac{\alpha^2(t)}{\Sigma} dt\right).$$

Rearrange the above equation as

$$\frac{\alpha^2(t)}{\left(\frac{1}{\Sigma_0+\sigma} + \int_0^t \frac{\alpha^2(t)}{\Sigma} dt\right)^2} = \left(1 + \frac{\sigma}{\Sigma_0}\right)^2\Sigma^2\lambda^2(t).$$

Integrating with respect to dt, we obtain

$$\alpha(t) = \frac{\lambda(t)}{\frac{\Sigma_0}{\Sigma} - \left(1 + \frac{\sigma}{\Sigma_0}\right)\int_0^t \lambda^2(t)dt}. \tag{8}$$

Since $\alpha(t) \geq 0$, $\lambda(t) > 0$ by assumption, we must have

$$\frac{\Sigma_0}{\Sigma} - \left(1 + \frac{\sigma}{\Sigma_0}\right)\int_0^t \lambda^2(t)dt > 0. \qquad t \in [0, 1). \tag{9}$$

This condition is crucial in the insider's profit optimization problem.

Now we can compute a.

$$
\begin{aligned}
a &\equiv \int_0^t \frac{\alpha^2(t)}{\Sigma} dt \\
&= \frac{1}{\Sigma} \int_0^t \frac{\lambda^2(t)}{\left(\frac{\Sigma_0}{\Sigma} - \left(1 + \frac{\sigma}{\Sigma_0}\right) \int_0^t \lambda^2(t)dt\right)^2} dt \\
&= \frac{1}{\Sigma}\left(1 + \frac{\sigma}{\Sigma_0}\right)^{-1} \left\{ \left[\frac{\Sigma_0}{\Sigma} - \left(1 + \frac{\sigma}{\Sigma_0}\right)\int_0^t \lambda^2(t)dt\right]^{-1} - \frac{\Sigma}{\Sigma_0} \right\} \\
&= \frac{\Sigma_0}{\Sigma(\Sigma_0 + \sigma)}\left[\frac{\Sigma_0}{\Sigma} - \left(1 + \frac{\sigma}{\Sigma_0}\right)\int_0^t \lambda^2(t)dt\right]^{-1} - \frac{1}{\Sigma_0 + \sigma}.
\end{aligned}
$$

The *ex ante* expected profit of the insider is

$$
\begin{aligned}
&\mathrm{E}\int_0^1 (\widetilde{v} - P(t))dX(t) \\
=\ &\mathrm{E}\int_0^1 (\widetilde{v} - P(t))[\alpha(t)\widetilde{s} - \beta(t)P(t)]dt \\
=\ &\mathrm{E}\int_0^1 (\widetilde{v} - P(t))[\alpha(t)\widetilde{s} - \beta(t)\widetilde{v} + \beta(t)(\widetilde{v} - P(t))]dt \\
=\ &\mathrm{E}\int_0^1 [\beta(t)(\widetilde{v} - P(t))^2 + \alpha(t)\widetilde{s}\widetilde{v} - \alpha(t)\widetilde{s}P(t) - \beta(t)\widetilde{v}^2 + \beta(t)\widetilde{v}P(t)]dt \\
=\ &\int_0^1 [\beta(t)\Gamma_{11}(t) + (\alpha(t) - \beta(t))\Sigma_0 - \alpha(t)f(t) + \beta(t)g(t)]dt.
\end{aligned}
$$

Here

$$
\begin{aligned}
\Gamma_{11}(t) &\equiv \mathrm{E}[\widetilde{v} - P(t)]^2 \\
&= \frac{\Sigma_0 \sigma}{1 + (\Sigma_0 + \sigma)a}(\sigma^{-1} + a) \\
&= \frac{\Sigma\Sigma_0}{\Sigma_0 + \sigma}\left(\frac{\Sigma_0}{\Sigma} - \left(1 + \frac{\sigma}{\Sigma_0}\right)\int_0^t \lambda^2(t)dt\right) + \frac{\Sigma_0\sigma}{\Sigma_0 + \sigma} \qquad (10)
\end{aligned}
$$

is the first component of the matrix $\Gamma(t)$, and

$$
f(t) = \mathrm{E}[\widetilde{s}P(t)], \qquad g(t) = \mathrm{E}[\widetilde{v}P(t)].
$$

From Eqs. (1) and (2), $f(t)$ and $g(t)$ are determined by

$$
\frac{df(t)}{dt} = \lambda(t)\alpha(t)(\Sigma_0 + \sigma) - \lambda(t)\beta(t)f(t), \qquad (11)
$$

$$
\frac{dg(t)}{dt} = \lambda(t)\alpha(t)\Sigma_0 - \lambda(t)\beta(t)g(t), \qquad (12)
$$

respectively. From Eqs. (11) and (12), we easily see that

$$
\frac{f(t)}{\Sigma_0 + \sigma} = \frac{g(t)}{\Sigma_0}.
$$

Thus by Eq. (7), we have $\alpha(t)f(t) = \beta(t)g(t)$.

Consequently, the ex ante expected profit of the insider is

$$
\begin{aligned}
\Pi(X,P) &= \int_0^1 [\beta(t)\Gamma_{11}(t) + (\alpha(t) - \beta(t))\Sigma_0 - \alpha(t)f(t) + \beta(t)g(t)]dt \\
&= \int_0^1 \left(1 + \frac{\sigma}{\Sigma_0}\right) \frac{\lambda(t)}{\frac{\Sigma_0}{\Sigma} - \left(1 + \frac{\sigma}{\Sigma_0}\right)\int_0^t \lambda^2(t)dt} \\
&\quad \times \left\{ \frac{\Sigma\Sigma_0}{\Sigma_0 + \sigma}\left(\frac{\Sigma_0}{\Sigma} - \left(1 + \frac{\sigma}{\Sigma_0}\right)\int_0^t \lambda^2(t)dt\right) + \frac{\Sigma_0\sigma}{\Sigma_0 + \sigma}\right\} dt \\
&\quad - \int_0^1 \frac{\sigma}{\Sigma_0}\frac{\Sigma_0\lambda(t)}{\frac{\Sigma_0}{\Sigma} - \left(1 + \frac{\sigma}{\Sigma_0}\right)\int_0^t \lambda^2(t)dt}dt \\
&= \int_0^1 \Sigma\lambda(t)dt. \qquad\qquad\qquad\qquad\qquad (13)
\end{aligned}
$$

We have to maximize the above quantity subject to the constraint given by Eq. (9).

By Schwarz inequality and Eq. (9), we have,

$$
\int_0^1 \lambda(t)dt \le \left(\int_0^1 \lambda^2(t)dt\right)^{\frac{1}{2}} \le \left(1 + \frac{\sigma}{\Sigma_0}\right)^{-\frac{1}{2}}\left(\frac{\Sigma_0}{\Sigma}\right)^{\frac{1}{2}}.
$$

The equality holds if and only if

$$
\lambda(t) = \left(1 + \frac{\sigma}{\Sigma_0}\right)^{-\frac{1}{2}}\left(\frac{\Sigma_0}{\Sigma}\right)^{\frac{1}{2}} = \frac{\Sigma_0}{\sqrt{\Sigma(\Sigma_0 + \sigma)}}, \qquad t \in [0,1). \qquad (14)
$$

Consequently, from Eqs. (8) and (7), we obtain

$$
\alpha(t) = \sqrt{\frac{\Sigma}{\Sigma_0 + \sigma}}\frac{1}{1-t}, \qquad \beta(t) = \frac{\sqrt{\Sigma(\Sigma_0 + \sigma)}}{\Sigma_0}\frac{1}{1-t}.
$$

This completes the proof.

COROLLARY 4.

$$
E[\tilde{v} - P(t)]^2 = \frac{\Sigma_0^2}{\Sigma_0 + \sigma}(1-t) + \frac{\Sigma_0\sigma}{\Sigma_0 + \sigma}.
$$

Proof. The conclusion follows from Eqs. (10) and (14).

We see that with the passage of time, more information is revealed, but it is never fully revealed since $\sigma > 0$.

COROLLARY 5. *In equilibrium, the ex ante expected profit of the insider is*

$$
\Pi(X,P) = \Sigma_0\sqrt{\frac{\Sigma}{\Sigma_0 + \sigma}}.
$$

Proof. The conclusion follows from Eqs. (13) and (14).

4. Conclusion

We have characterized the linear Nash equilibrium in a generalized Kyle's dynamic insider trading model, and have quantified the incorporation of private information into the equilibrium price within a rational expectations framework. Compared with the static case in Kyle (1985), we find that the dynamic trading improves the profit of the insider since he has an additional dimension to timing his trading strategies so that he can exploits more thoroughly his private information. The single insider acts as an "information monopolist" by attempting to extract the most profit from his superior private information. The absence of competition gives the insider a great flexibility regarding the timing of his trades. If, instead, there are multiple insiders with diverse private information, and they act competitively and strategically, then a richer pattern of information revelation and welfare effect is expected. This case is considerably more complicated and we will not pursue it here.

References

1. A. R. Admati, A noise rational expectations equilibrium for multi-asset securities markets, *Econometrica*, 53 (1985), 629-657.
2. A. R. Admati, P. Pflerderer, A theory of intraday patterns: Volume, and price variability, *Review of Financial Studies*, 1 (1988), 3-40.
3. K. Back, Insider trading in continuous time, *Review of Financial Studies*, 5 (1992), 387-409.
4. M. J. Brennan, H. H. Cao, Information, trade, and the derivative securities, *Review of Financial Studies*, 9 (1996), 163-208.
5. J. Y. Cambel, A. Kyle, Smart money, noise trading and stock price behavior, *Review of Economic Studies*, 60 (1993), 1-34.
6. D. W. Diamond, R. E. Verrecchia, Information aggregation in a noise rational expectations economy, *Journal of Financial Economy*, 9 (1981), 221-235.
7. S. J. Grossman, J. E. Stiglitz, On the impossibility of informationly efficient markets, *American Economic Review*, 70 (1980), 393-408.
8. M. F. Hellwig, On the aggregation of information in competitive market, *Journal of Economic Theory*, 22 (1980), 477-498.
9. C. W. Holden, A. Subrahmanyam, Long-lived information and imperfect competition, *Journal of Finance*, 47 (1992), 247-270.
10. A. Kyle, Continuous auctions and insider trading, *Econometrica*, 53 (1985), 1315-1335.
11. A. Kyle, Informed speculation with imperfect competition, *Review of Economic Studies*, 56 (1989), 317-356.
12. R. S. Lipster, A. N. Shirayaev, *Statistics of Random Processes*, I. Springer-Verlag, Berlin, 1977.
13. M. O'Hara, *Market Microstructure Theory*, Blackwell, Cambridge, 1995.
14. J. C. Rochet, J. L. Vila, Insider trading without normality, *Review of Economic Studies*, 61 (1994), 131-152.
15. M. Spiegel, A. Subrahmanyam, Informed speculation and hedging in a noncompetitive securities markets, *Review of Financial Studies*, 5 (1992), 307-329.
16. R. E. Verrecchia, Information acquisition in a noisy rational expectations economy, *Econometrica*, 50 (1982), 1415-1432.
17. J. Wang, A model of intertemporal asset prices under asymmetric information, *Review of Economic Studies*, 60 (1993), 269-282.

Institute of Applied Mathematics, Academia Sinica, Beijing, 100080, P. R. China,
E-mail: luoslamath4.amt.ac.cn,

Department of Economics and Finance, City University of Hong Kong, Tat Chee
Avenue, Kowloon, Hong Kong, and Department of Applied Mathematics and Statistics,
State University of New York at Stony Brook, Stony Brook, New York, 11794-3600,
USA

AMS/IP Studies in Advanced Mathematics
Volume 26, 2002

A New Hedging Model and a Nonlinear Generalization of Black-Scholes Formula

Shanjian Tang

ABSTRACT. In a financial market, there are two types of information — technical and fundamental. Those, like the past and current prices of securities, which are used in technical analyses, are called technical information. While those which are used in fundamental analyses, like the financial reports and the potential developing possibilities on the enterprises with which the underlying securities are associated, are called fundamental information. Both types of information are concerned and used by financial decision-makers like investors and hedgers. However, most (if not all) of the existing mathematical models for the analysis of investing and pricing take into consideration the technical information only, and no parameters which directly indicate the underlying fundamental economic situation go into those models. So, all the analyses based on those models are technical ones. In this paper, a new stochastic control model is presented for a hedger, which incorporates not only the technical information but also the fundamental information. Using the stochastic optimal control theory, a nonlinear generalized version of Black-Scholes formula is obtained for the European option. Actually, a generalized pricing formula for a general contingent claim is given as the solution of a linear backward stochastic differential equation. These new formulas depend upon a nonlinear Riccati differential equation.

1. Introduction

A control theorist looks at a financial market in the following way. A hedger is a controller, and his/her controlling tool is his/her portfolio. The state he/she wants to control is his/her wealth level, that is, the total market value of all his/her securities. The dynamics of the state — the wealth level — depends on the dynamics of all security prices in the market and his portfolio as well. Assume that the prices of all the securities in the financial markets evolve according to linear stochastic differential equations (SDE in short). Then, the hedger's wealth level also satisfies a linear SDE, whose coefficients depend on his/her portfolio. One trade by the hedger — one change of his/her portfolio position — means that he/she is imposing one control action on his wealth level. Obviously, different ways of trading by the hedger will give rise to different dynamics of his/her wealth level. The hedger's task is to trade securities — to choose his/her portfolio (control) — carefully to approach

Project 79790130 supported by **the National Natural Science Foundation of China**.

his/her desired contingent claim with the least "risk". This viewpoint from the optimal control theory is developed in what follows.

1.1. The Black-Scholes financial market model. Let $W = (W_1, \ldots, W_d)^*$ be a d-dimensional Wiener process, defined on a complete probability space $(\Omega, \mathcal{F}, P)$, and let $\{\mathcal{F}_t, 0 \leq t \leq T\}$ be the P-augmentation of the natural filtration generated by the Wiener process W. Here and in the following, the asterisk in the superscript denotes the transpose.

Consider a financial market in which there are $n + 1$ securities on the time interval $[0, T]$. One of them is non-risky, called the bond. Its price at time t is denoted by $B(t)$. It evolves according to the following ordinary differential equation

$$(1) \qquad \begin{cases} dB(t) & = \quad r(t)B(t)\,dt, \quad 0 < t \leq T, \\ B(0) & = \quad 1 \end{cases}$$

where $r(t)$ is *the instantaneous interest rate* of the bond at time t and is assume to be deterministic and essentially bounded. The other n securities are risky, called stocks. The price for one share of the ith stock is denoted by $X_i(t)$. The price vector process $X = (X_1, \cdots, X_n)^*$ is modelled by the linear SDE

$$(2) \qquad \begin{cases} dX_i(t) & = \quad X_i(t)\left[b_i(t)\,dt + \displaystyle\sum_{j=1}^{d} \sigma_{ij}(t)\,dW_j(t) \right], \quad 0 < t \leq T, \\ X_i(0) & = \quad x_i. \end{cases}$$

Set

$$\sigma_j =: (\sigma_{1j}, \ldots, \sigma_{nj})^*, \quad \sigma = (\sigma_1, \cdots, \sigma_d), \quad b =: (b_1, \ldots, b_n)^*.$$

b and σ are interpreted as *the appreciation rate vector* and *the volatility matrix*. They are assumed to be deterministic and essentially bounded.

1.2. Portfolio processes and wealth processes of hedgers. Consider a hedger who invests in the $n + 1$ basic securities. At time 0, he invests the amount $v \in \mathcal{R}$ in the $n+1$ securities. Denote by $V(t)$ the market value of the amount which he invests at time t in the $n + 1$ securities. Then,

$$V(0) = v.$$

For $i = 1, \ldots, d$, denote by $\pi_i(t)$ the amount that he invests in the ith stock at time t. The amount that he invests in the bond at time t is then given by

$$V(t) - \sum_{i=1}^{n} \pi_i(t).$$

Let $\mathcal{A}(t, T)$ be the totality of $\mathcal{R}^n$-valued $\mathcal{F}_t$-adapted processes

$$\{\pi(s) = (\pi_1(s), \ldots, \pi_n(s))^*; t \leq s \leq T\}$$

which satisfies

$$E \int_t^T |\pi(s)|^2\,ds < \infty.$$

The hedger's portfolio process $\{(\pi_1(s), \cdots, \pi_n(s), V(s) - \sum_{i=1}^{n} \pi_i(s))^*; t \leq s \leq T\}$ is said to be *self-financing*, if it satisfies the following SDE

$$(3) \quad dV(t) = \sum_{i=1}^{n} \pi_i(t) \left[b_i(t) \, dt \right.$$

$$(4) \qquad \left. + \sum_{j=1}^{d} \sigma_{ij}(t) dW_j(t) \right] + \left(V(t) - \sum_{i=1}^{n} \pi_i(t) \right) r(t) \, dt, 0 < t \leq T.$$

For each $\pi \in \mathcal{A}(t,T)$, the hedger's portfolio process $(\pi^*, V - \sum_{i=1}^{n} \pi_i)^*$ is always assumed to be self-financing, or equivalently to say, the hedger's wealth V is always assumed to satisfy the SDE (4). So, (4) is called the hedger's *wealth equation*.

Let $\mathbf{1}$ be the n-dimensional column vector whose every component is 1. Then the above equation is equivalently written as follows:

$$(5) \quad dV(t) = r(t)V(t) \, dt + \tilde{b}^*(t)\pi(t) \, dt + \pi^*(t)\sigma(t) \, dW(t), \quad 0 < t \leq T$$

with

$$(6) \qquad \tilde{b}(t) := b(t) - r(t)\mathbf{1}, \quad 0 \leq t \leq T.$$

1.3. Behavior of hedgers: the risk-minimizing principle. To approach the given contingent claim $\xi \in L^2(\Omega, \mathcal{F}_T, P)$, the hedger chooses his trading policy so as to minimize, over $\pi \in \mathcal{A}(t,T)$, the following risk functional

$$(7) \qquad J(\pi) = E(V(T) - \xi)^2 + E \int_0^T \pi^*(t) R(t) \pi(t) \, dt.$$

The first term in (7) represents the risk of hedging loss. The novelty of this paper is that we introduce a new term $R(\cdot)$ — called the hedger's risk metric tensor fields which carries the hedger's fundamental information about the economic situation behind the market — into the risk functional (7) which is to be minimized by the hedger. $R(\tau)$ can be interpreted as a measure by the hedger on the "credit" risk of the underlying financial market and/or the associated financial institutions, coming from the uncertainty of the model (unavoidable modelling errors, for example), some unexpected crucial events, and so on.

The author would like to thank David Heath for providing the following comment in the course of the workshop. Assume that the modelling errors $\delta r, \delta b$ and $\delta \sigma$ of the market conditions r, b and σ are Gaussian processes. Then it might produce the second term of (7) to calculate the expected square of the hedging loss. Due to the limitation of space, I would like to leave this idea developed elsewhere in details.

1.4. Comments. Optimal stochastic control has been being practised by investors since the birth of security markets. The application of the optimal stochastic control theory to financial economics is dated back at least to the late 1960s' works of Samuelson and Merton, and the reader is referred to the book [**18**] and the references therein. Since then this application has produced fruitful results. For these applications and the related works, the reader is referred to among others, Barron and Jensen [**1**], Bertsekas [**2**], Bismut [**3**], Cox and Huang [**4**], Cvitanic and Karatzas [**5, 6**], Duffie and Richardson [**8**], El Karoui and Quenez [**9**], Föllmer and Schweizer [**11**], Karatzas [**12**], Karatzas, Lehoczky, and Shreve [**13**], Karatzas, Lehoczky, Shreve, and Xu [**14**], Kohlmann and Tang [**15**], Kohlmann and Zhou [**16**],

Kwok [17], Richardson [24], Schweizer [25], Tang [26], and Zhou and Li [27]. In these applications, the theory of backward stochastic differential equations often plays an important role, and the reader is referred to Pardoux and Peng [19], Duffie and Epstein [7], and El Karoui, Peng, and Quenez [10] for that theory.

However, in most (if not all) of existing models for investment, consumption and pricing, the portfolio (the control variable) does not enter into the objective functional explicitly. Here we propose to study a generalization of Duffie and Richardson's mean-variance hedging model (see Duffie and Richardson [8]) by introducing the portfolio into the objective functional, and to investigate the resulting new optimal hedging process.

Existing models for the analysis of option pricing and investment/consumption problems, strongly depend upon the hypothesis that the volatility coefficients are not degenerate, such as those models in El Karoui and Quenez [9] and Cvitanic and Karatzas [5, 6]. Kohlmann and Tang [15] showed that this hypothesis is crucial to guarantee the existence and uniqueness of the optimal hedging policy for the mean-variance hedging problem. The hypothesis of nondegeneracy is a fatal disadvantage, and makes it not applicable to financial markets with possible zero volatility. When the volatility becomes zero, the hedger with $R \equiv 0$ will trade in a mad way. This is contrast to the hedger with $R > 0$. It is known that, if $R(t)$ is assumed to be positive for $\forall t$, then it is not necessary to assume that $\sigma\sigma^*$ is positive. Thus the framework presented here essentially applies to incomplete markets. In the following, for convenience of simplicity, R is assumed to be uniformly positive. Note that R can also be negative if $\sigma\sigma^*$ is sufficiently positive. Some lower negative bound for R, given the lower positive bound of $\sigma\sigma^*$, can be given by refining the relevant arguments of Kohlmann and Tang [15].

The rest of the paper is organized as follows. In Section 2, a general principle to price a contingent claim via the mean-variance hedging is described. In Sections 3 and 4, the problems of pricing a futures contract and a European option are solved respectively, and the explicit pricing formulas are given, which are associated with the solution of a nonlinear Riccati differential equation. When R is set to zero, that Riccati equation becomes linear, and thus its solution is explicitly solvable. So, for general nonzero R, the pricing formula obtained here for the European option is a *nonlinear* generalization of the well-known Black-Scholes formula. In Section 5, the pricing formula is derived for a general contingent claim as the solution of a backward stochastic differential equation, which is again associated with the solution of a nonlinear Riccati differential equation. Finally, in Section 6, a remark is given on the case where the market conditions (the instantaneous interest rate, the appreciation rate vector, and the volatility matrix) are allowed to be random.

2. Mechanism for Pricing a Contingent Claim via the Mean-Variance Hedging

Let $\xi \in \mathcal{L}^2(\Omega, \mathcal{F}_T, P)$ be a contingent claim at time T. Consider the following risk functional at time t

$$(8) \qquad J_{t,v}(\pi) = E\left[\left(V^{t,v}(T) - \xi\right)^2 + \int_t^T \pi^*(\tau) R(\tau) \pi(\tau)\, d\tau \,\Big|\, \mathcal{F}_t \right].$$

Here, $V^{t,v}(\cdot)$ solves the following SDE

(9)
$$\begin{cases} dV(\tau) & = & r(\tau)V(\tau)\,d\tau + \tilde{b}^*(\tau)\pi(\tau)\,d\tau + \pi^*(\tau)\sigma(\tau)\,dW(\tau), \quad t < \tau \le T, \\ V(t) & = & v \in \mathcal{R}. \end{cases}$$

Set

(10)
$$u(t,v) := \inf_{\pi \in \mathcal{A}(t,T)} J_{t,v}(\pi).$$

The minimizing $\pi(\cdot) \in \mathcal{A}(t,T)$ is called the hedger's *optimal hedging process* with the wealth v at the initial time t.

If there is $\bar{v} \in \mathcal{L}^2(\Omega, \mathcal{F}_t, P)$ such that

(11)
$$u(t,\bar{v}) = \min_{v \in \mathcal{R}} u(t,v), \quad a.s.$$

then $\bar{v}$ is called *the hedger's subjective price* on the contingent claim ξ at time t. An $\mathcal{F}_t$-adapted stochastic continuous process $v(\cdot)$ is called *the hedger's subjective price process* on the contingent claim ξ if $v(t)$ is the hedger's subjective price on the contingent claim ξ at time t for $\forall t \in [0,T]$.

In particular, consider the case of $\xi = g(X_1(T), \cdots, X_n(T))$ for some continuous function $g : \mathcal{R}^n \to \mathcal{R}$. Then, consider the following optimal control problem. Treat both the security price processes $X_1, \ldots, X_n$ and the wealth process V as states. The cost functional is

(12)
$$J_{t,x,v}(\pi) = E\left(V^{t,v}(T) - g(X_1^{t,x}(T), \cdots, X_n^{t,x}(T))\right)^2 + E\int_t^T \pi^*(\tau)R(\tau)\pi(\tau)\,d\tau$$

where $V^{t,v}(\cdot)$ is the solution to (9) and $X_1^{t,x}, \ldots, X_n^{t,x}$ solve respectively the following SDEs

(13)
$$\begin{cases} dX_i(s) & = & X_i(s)\left[b_i(s)\,ds + \displaystyle\sum_{j=1}^d \sigma_{ij}(s)\,dW_j(s)\right], \quad t < s \le T, \\ X_i(t) & = & x_i, \quad x = (x_1, \cdots, x_n)^* \in \mathcal{R}^n. \end{cases}$$

The value function is

(14)
$$U(t,x,v) =: \inf_{\pi \in \mathcal{A}(t,T)} J_{t,x,v}(\pi).$$

The minimizing $\bar{\pi}(t,x,v), (t,x) \in [0,T] \times \mathcal{R}^n$, in closed form is called the hedger's *optimal hedging policy* with the wealth v at time t. If $\bar{v} : [0,T] \times \mathcal{R}^n \to \mathcal{R}$ satisfies

(15)
$$U(t,x,\bar{v}(t,x)) = \min_{v \in \mathcal{R}} U(t,x,v), \quad \forall (t,x),$$

then $\bar{v}$ is called *the hedger's subjective price function* on the contingent claim ξ, and $\bar{\pi}(t,x,\bar{v}(t,x)), (t,x) \in [0,T] \times \mathcal{R}^n$, is called the hedger's *doublely optimal hedging policy*.

If further the values of the process $\pi(\cdot) \in \mathcal{A}(0,T)$ are constrainted within a given subset $S \subset \mathcal{R}^n$, then the value function U should satisfy the following

Hamilton-Jacobi-Bellman (HJB in short) equation

(16)

$$
\begin{cases}
U_t + \dfrac{1}{2} \displaystyle\sum_{i,j=1}^{n} \sum_{k=1}^{d} \sigma_{ik} x_i \sigma_{jk} x_j U_{x_i x_j} + \sum_{i=1}^{n} b_i x_i U_{x_i} + rv U_v \\[2ex]
\quad + \displaystyle\inf_{\pi \in S} \left\{ \dfrac{1}{2} \sum_{k=1}^{d} (\sigma_k^* \pi)^2 U_{vv} + \tilde{b}^* \pi U_v + \sum_{i=1}^{n} \sum_{k=1}^{d} \sigma_{ik} x_i \sigma_k^* \pi U_{x_i v} + \pi^* R(t) \pi \right\} = 0, \\[2ex]
U(T, x, v) = (v - g(x))^2 .
\end{cases}
$$

3. Pricing a Futures Contract

Consider the case of $n = 1$, i.e., there is only one risky security in the market. For a futures contract for one share of the stock with the striking price K and the expiration time T, the contingent claim $\xi = X(T) - K$. Set $|\sigma|^2 = \sigma_1^2 + \cdots + \sigma_d^2$.

Set

$$
Y(t) = (X(t), V(t))^*, \quad t \in [0, T].
$$

Then $Y(\cdot)$ satisfies

(17)

$$
\begin{cases}
dY(t) & = & [A(t) Y(t) + B(t)\pi(t)]\, dt + \displaystyle\sum_{j=1}^{d} [C_j(t) Y(t) + D_j(t)\pi(t)]\, dW_j(t), \\[2ex]
Y(0) & = & (x, v)^*
\end{cases}
$$

where

$$
A(t) \quad := \quad \begin{bmatrix} b(t) & 0 \\ 0 & r(t) \end{bmatrix},
$$

(18)

$$
B(t) \quad := \quad \begin{bmatrix} 0 \\ \tilde{b}(t) \end{bmatrix},
$$

$$
C_j(t) \quad := \quad \begin{bmatrix} \sigma_j(t) & 0 \\ 0 & 0 \end{bmatrix}, \quad D_j(t) := \begin{bmatrix} 0 \\ \sigma_j(t) \end{bmatrix}, \quad j = 1, \ldots, d.
$$

The risk functional reads

(19)

$$
\begin{aligned}
J(\pi)_{t,x,v} & = \frac{1}{2} E \left\{ Y^*(T) \begin{bmatrix} 1 & -1 \\ -1 & 1 \end{bmatrix} Y(T) \right\} + E \left\{ Y^*(T) \begin{bmatrix} -K \\ K \end{bmatrix} \right\} \\[2ex]
& \quad + \frac{1}{2} E \int_0^T \pi^*(t) R(t) \pi(t)\, dt.
\end{aligned}
$$

The associated differential equations are

$$(20) \quad \begin{cases} \begin{aligned} \frac{d}{dt}Q(t) &= -A^*(t)Q(t) - Q(t)A(t) - \sum_{j=1}^{d} C_j^*(t)Q(t)C_j(t) \\ &\quad - \left[Q(t)B(t) + \sum_{j=1}^{d} C_j(t)Q(t)D_j(t) \right] \\ &\quad \times \left(R(t) + \sum_{j=1}^{d} D_j^*(t)P(t)D_j(t) \right)^{-1} \\ &\quad \times \left[B^*(t)Q(t) + \sum_{j=1}^{d} D_j^*(t)Q(t)C_j^*(t) \right], \\ Q(T) &= \begin{bmatrix} 1 & -1 \\ -1 & 1 \end{bmatrix} \end{aligned} \end{cases}$$

and

$$(21) \quad \begin{cases} \begin{aligned} \frac{d}{dt}p(t) &= -A(t)p(t) + \left[Q(t)B(t) + \sum_{j=1}^{d} C_j(t)Q(t)D_j(t) \right] \\ &\quad \times \left[R(t) + |\sigma|^2 Q_{22}(t) \right]^{-1} B^*(t)p(t) \\ &= -\begin{bmatrix} b(t) & -\dfrac{\tilde{b}(\tilde{b} + |\sigma|^2)Q_{12}(t)}{R(t) + |\sigma|^2 Q_{22}(t)} \\ 0 & r(t) - \dfrac{\tilde{b}^2(t)Q_{22}(t)}{R(t) + |\sigma|^2(t)Q_{22}(t)} \end{bmatrix} p(t), \quad 0 \le t < T, \\ p(T) &= (-K, K)^*. \end{aligned} \end{cases}$$

Here and in the following, write

$$Q(t) := \begin{bmatrix} Q_{11}(t) & Q_{12}(t) \\ Q_{12}^*(t) & Q_{22}(t) \end{bmatrix}, \quad p(t) := \begin{pmatrix} p_1(t) \\ p_2(t) \end{pmatrix}, \quad 0 \le t < T.$$

It follows that

$$(22) \quad \begin{cases} \dfrac{d}{dt}Q_{11}(t) &= -2bQ_{11}(t) - |\sigma|^2 Q_{11}(t) + \dfrac{(\tilde{b} + |\sigma|^2)^2}{R + |\sigma|^2 Q_{22}} Q_{12}^2(t), \\ Q_{11}(T) &= 1; \end{cases}$$

$$(23) \quad \begin{cases} \dfrac{d}{dt}Q_{12}(t) &= -(b + r)Q_{12}(t) + \dfrac{\tilde{b}(\tilde{b} + |\sigma|^2)Q_{22}}{R + |\sigma|^2 Q_{22}} Q_{12}(t), \\ Q_{12}(T) &= -1; \end{cases}$$

$$(24) \quad \begin{cases} \dfrac{d}{dt}Q_{22}(t) &= -2rQ_{22}(t) + \dfrac{\tilde{b}^2 Q_{22}^2(t)}{R + |\sigma|^2 Q_{22}(t)}, \\ Q_{22}(T) &= 1. \end{cases}$$

The value function is given by

$$(25) \qquad U(t,x,v) = \frac{1}{2}Q_{11}(t)x^2 + \frac{1}{2}Q_{22}(t)v^2 + Q_{12}(t)xv + p_1(t)x + p_2(t)v.$$

Therefore, the optimal hedging policy $\bar{\pi}$ is

$$(26) \qquad \bar{\pi}(t,x,v) = -\frac{(\tilde{b}+|\sigma|^2)Q_{12}x + \tilde{b}Q_{22}v + \tilde{b}p_2}{R + |\sigma|^2 Q_{22}}.$$

The hedger's subjective price function $\bar{v}$ on the futures contract is determined by

$$(27) \qquad U(t,x,\bar{v}) - \min_{v\in\mathcal{R}} U(t,x,v)$$

which implies immediately

$$(28) \qquad \bar{v}(t,x) = -\frac{p_2(t) + Q_{12}(t)x}{Q_{22}(t)}.$$

While

$$(29) \qquad \begin{aligned} \frac{p_2(t)}{Q_{22}(t)} &= K\exp\left(-\int_t^T r(s)\,ds\right), \\ \frac{Q_{12}(t)}{Q_{22}(t)} &= -\exp\left(\int_t^T \frac{R(s)\tilde{b}(s)}{R(s) + |\sigma|^2 Q_{22}(s)}\,ds\right), \end{aligned}$$

therefore, *the hedger's subjective price function* has the following explicit formula

$$(30) \qquad \bar{v}(t,x) = x\exp\left(\int_t^T \frac{R(s)\tilde{b}(s)}{R(s)+|\sigma|^2 Q_{22}(s)}\,ds\right) - K\exp\left(-\int_t^T r(s)\,ds\right).$$

Hence, *the doublely optimal hedging policy* is given by

$$(31) \qquad \begin{aligned} \bar{\pi}(t,x,\bar{v}(t,x)) &= -\frac{|\sigma|^2 Q_{12}(t)}{R(t)+|\sigma|^2 Q_{22}(t)}x \\ &= \frac{|\sigma|^2 Q_{22}(t)}{R(t)+|\sigma|^2 Q_{22}(t)}x\exp\left(\int_t^T \frac{R(s)\tilde{b}(s)}{R(s)+|\sigma|^2 Q_{22}(s)}\,ds\right). \end{aligned}$$

4. Pricing a European Option

Consider the case of $n = 1$, i.e., there is only one risky security in the market. In the case of a European call option for one share of stock with the striking price K and the expirate time T, the contingent claim $\xi = (X(T) - K)^+$.

The value function should satisfy the following HJB equation

$$(32) \qquad \begin{cases} U_t + \dfrac{1}{2}|\sigma|^2 x^2 U_{xx} + bx U_x + rv U_v \\[2mm] \quad + \min_{\pi\in\mathcal{R}}\left\{\dfrac{1}{2}|\sigma|^2\pi^2 U_{vv} + \tilde{b}\pi U_v + |\sigma|^2\pi x U_{xv} + R\pi^2\right\} = 0, \\[2mm] U(T,x,v) = [v - (x-K)^+]^2. \end{cases}$$

Since

(33)

$$\min_{\pi \in \mathcal{R}} \left\{ \frac{1}{2} |\sigma|^2 \pi^2 U_{vv} + \tilde{b}\pi U_v + |\sigma|^2 \pi x U_{xv} + R\pi^2 \right\} = -\frac{1}{2} \frac{\left(\tilde{b}U_v + |\sigma|^2 x U_{xv} \right)^2}{2R + |\sigma|^2 U_{vv}},$$

with the minimizing point (which gives *the hedger's optimal hedging policy*) being

(34)
$$\bar{\pi}(t, x, v) = -\frac{\tilde{b}U_v + |\sigma|^2 x U_{xv}}{2R + |\sigma|^2 U_{vv}},$$

the equation (32) is equivalent to

(35)
$$\begin{cases} U_t + \frac{1}{2} |\sigma|^2 x^2 U_{xx} + bx U_x + rv U_v - \frac{1}{2} \frac{\left(\tilde{b}U_v + |\sigma|^2 x U_{xv} \right)^2}{2R + |\sigma|^2 U_{vv}} = 0, \\ \\ U(T, x, v) = [v - (x - K)^+]^2. \end{cases}$$

Suppose that (35) has a solution of the following form

(36) $U(t, x, v) = [v^2 - 2v\bar{v}(t, x)]Q_{22}(t) + \hat{v}(t, x), \quad t \in [0, T], x > 0, v \in \mathcal{R}.$

Here $\bar{v}$ and $\hat{v}$ are deterministic functions of (t, x), waiting to be determined later. The following is verified directly.

$$\begin{aligned} U_t &= -\left(2r - \frac{\tilde{b}^2 Q_{22}}{R + |\sigma|^2 Q_{22}} \right) Q_{22}[v^2 - 2v\bar{v}(t, x)] \\ &\quad -2Q_{22}v\bar{v}_t(t, x) + \hat{v}_t(t, x), \end{aligned}$$

(37)
$$\begin{aligned} U_{xx} &= -2Q_{22}v\bar{v}_{xx}(t, x) + \hat{v}_{xx}(t, x), \\ U_x &= -2Q_{22}v\bar{v}_x(t, x) + \hat{v}_x(t, x), \\ U_v &= 2Q_{22}[v - \bar{v}(t, x)], \\ U_{vv} &= 2Q_{22}, \\ U_{xv} &= -2Q_{22}\bar{v}_x(t, x). \end{aligned}$$

Then,

$$U_t + \frac{1}{2} |\sigma|^2 x^2 U_{xx} + bx U_x + rv U_v$$

(38)
$$= -2v \left(\bar{v}_t + \frac{1}{2} |\sigma|^2 x^2 \bar{v}_{xx} + rx\bar{v}_x - r\bar{v} + \tilde{b}x\bar{v}_x + \frac{\tilde{b}^2 Q_{22}}{R + |\sigma|^2 Q_{22}} \bar{v} \right) Q_{22}$$

$$+ \frac{\tilde{b}^2 Q_{22}^2}{R + |\sigma|^2 Q_{22}} v^2 + \hat{v}_t + \frac{1}{2} |\sigma|^2 x^2 \hat{v}_{xx} + bx\hat{v}_x,$$

(39)
$$\frac{1}{2} \frac{\left(\tilde{b}U_v + |\sigma|^2 x U_{xv} \right)^2}{2R + |\sigma|^2 U_{vv}} = \frac{\tilde{b}^2 Q_{22}^2}{R + |\sigma|^2 Q_{22}} v^2 - \frac{2\tilde{b}Q_{22}^2}{R + |\sigma|^2 Q_{22}} v(\tilde{b}\bar{v} + |\sigma|^2 x\bar{v}_x)$$

$$+ \frac{Q_{22}}{R + |\sigma|^2 Q_{22}^2} (\tilde{b}\bar{v} + |\sigma|^2 x\bar{v}_x)^2,$$

$$U_t + \frac{1}{2}|\sigma|^2 x^2 U_{xx} + bx U_x + rv U_v - \frac{1}{2} \frac{\left(\tilde{b}U_v + |\sigma|^2 x U_{xv}\right)^2}{2R + |\sigma|^2 U_{vv}}$$

(40)

$$= -2v\left[\bar{v}_t + \frac{1}{2}|\sigma|^2 x^2 \bar{v}_{xx} + \left(r + \frac{R\tilde{b}}{R + |\sigma|^2 Q_{22}}\right) x\bar{v}_x - r\bar{v}\right] Q_{22}$$

$$+\hat{v}_t + \frac{1}{2}|\sigma|^2 x^2 \hat{v}_{xx} + bx\hat{v}_x - \frac{Q_{22}^2}{R + |\sigma|^2 Q_{22}}(\tilde{b}\bar{v} + |\sigma|^2 x \bar{v}_x)^2.$$

Let $\bar{v}$ satisfy the following partial differential equation

(41)

$$\begin{cases} \bar{v}_t + \frac{1}{2}|\sigma|^2 x^2 \bar{v}_{xx} + \left(r + \frac{R\tilde{b}}{R + |\sigma|^2 Q_{22}}\right) x\bar{v}_x = r\bar{v}, \\[2mm] \bar{v}(T,x) = (x - K)^+ \end{cases}$$

and let $\hat{v}$ satisfy the following partial differential equation

(42)

$$\begin{cases} \hat{v}_t + \frac{1}{2}|\sigma|^2 x^2 \hat{v}_{xx} + bx\hat{v}_x - \frac{Q_{22}^2}{R + |\sigma|^2 Q_{22}}(\tilde{b}\bar{v} + |\sigma|^2 x \bar{v}_x)^2 = 0, \\[2mm] \hat{v}(T,x) = \left[(x - K)^+\right]^2. \end{cases}$$

At this stage, it follows that assumption (36) is true. Then, $\bar{v}$ is *the hedger's subjective price function* on the call option, which is characterized by the partial differential equation (41). That equation is obviously a generalized version of the Black-Scholes equation (41) with $R \equiv 0$. Its solution has the following explicit formula

(43)

$$\bar{v}(t,x) = x \exp\left(\int_t^T \frac{R(s)\tilde{b}(s)}{R(s) + |\sigma|^2 Q_{22}(s)} \, ds\right) N(d_+)$$

$$-K \exp\left(-\int_t^T r(s)\, ds\right) N(d_-)$$

where

(44)

$$d_\pm = \frac{\log\left(\frac{x}{K}\right) + \int_t^T \left(r(s) + \frac{R(s)\tilde{b}(s)}{R(s) + |\sigma|^2(s)Q_{22}(s)} \pm \frac{|\sigma|^2}{2}\right) ds}{\sqrt{\int_t^T |\sigma|^2(s)\, ds}},$$

$$N(y) = \frac{1}{\sqrt{2\pi}} \int_{-\infty}^y \exp\left(-\frac{x^2}{2}\right) dx.$$

When $R \equiv 0$, Q_{22} disappears and the above formula becomes Black-Scholes formula. So, the formula (43) is a generalized version of Black-Scholes formula. Since it is associated with the nonlinear Riccati differential equation (24), it is a nonlinear genealized version.

The hedger's doublely optimal hedging policy is derived from (34) as

(45) $$\bar{\pi}(t, x, \bar{v}(t, x)) = \frac{|\sigma|^2 Q_{22}(t)\bar{v}_x(t, x)}{R(t) + |\sigma|^2 Q_{22}(t)} x.$$

5. Pricing a General Contingent Claim

Consider the general case posed in Section 2. We have

Theorem 1. 1. *The value function u has the following form*

(46) $$u(t, v) = [v^2 - 2vp(t)]Q(t) + \widehat{p}(t), \quad t \in [0, T], v \in \mathcal{R}$$

where $p(\cdot)$ and $\widehat{p}(\cdot)$ are characterized by the following linear backward stochastic differential equation

(47)
$$\begin{cases} dp(s) = \left[rp(s) + \tilde{b}^*(R + Q\sigma\sigma^*)^{-1}Q\sigma q(s)\right] ds + q^*(s)\, dW(s), \quad s \in [t, T), \\ p(T) = \xi \end{cases}$$

and the formula

(48)
$$\widehat{p}(t) := -E\left[\int_t^T (Qp\tilde{b} + Q\sigma q)^*(R + Q\sigma\sigma^*)^{-1}(Qp\tilde{b} + Q\sigma q)\, ds \,\bigg|\, \mathcal{F}_t\right] + E\left[\xi^2 | \mathcal{F}_t\right],$$

respectively, and $Q(\cdot)$ solves the following deterministic nonlinear Riccati differential equation

(49)
$$\begin{cases} \dfrac{d}{dt}Q(t) = -2rQ(t) + \tilde{b}^*\,(R + \sigma\sigma^*Q)^{-1}\,\tilde{b}Q^2(t), \quad t \in [0, T), \\ Q(T) = 1. \end{cases}$$

2. *The hedger's subjective price process on the contingent claim ξ is $p(\cdot)$, and the optimal hedging process is*

(50) $$\pi(s) = -(R + Q\sigma\sigma^*)^{-1}Q[V\tilde{b}(s) - p\tilde{b}(s) - \sigma q(s)], \quad s \in [t, T].$$

Proof The first assertion is concluded from the following observation: $u(\cdot, \cdot)$ is the value function of the following one-dimensional linear-quadratic optimal stochastic control problem: to minimize

$$E(V(T) - \xi)^2 + E\int_t^T \pi^*(s)R(s)\pi(s)\, ds$$

subject to the following dynamics

(51)
$$\begin{cases} dV(s) = [rV(s) + \tilde{b}^*\pi(s)]\, ds + \pi^*\sigma(s)\, dW(s), \quad s \in (t, T], \\ V(t) = v \end{cases}$$

and the corresponding Riccati's equation is exactly (49). Note that here and in the following, E is used instead of $E[\cdot|\mathcal{F}_t]$ for simplification of notation.

In fact, we have

(52) $$J_{t,v}(\pi) = E\left(V^2(T)\right) - 2E(V(T)\xi) + E(\xi^2) + E\int_t^T \pi^*(s)R\pi(s)\, ds,$$

while using Itô's formula we have

(53)
$$\begin{cases} dV^2(s) &= [2rV^2(s) + \pi^*\sigma\sigma^*\pi(s) + 2V\tilde{b}^*\pi(s)]\,ds \\ &\quad + 2V\pi^*\sigma(s)\,dW(s), \quad s \in (t,T], \\ V^2(t) &= v^2 \end{cases}$$

and

(54)
$$\begin{cases} d[V^2Q(s)] &= Q(s)[\pi^*\sigma\sigma^*\pi(s) + 2V\tilde{b}^*\pi(s) + \tilde{b}^*(R + \sigma\sigma^*Q)^{-1}\tilde{b}QV^2(s)]\,ds \\ &\quad + 2QV\pi^*\sigma(s)\,dW(s), \quad s \in (t,T], \\ V^2Q(t) &= v^2Q(t), \quad V^2Q(T) = V^2(T). \end{cases}$$

Therefore, it follows from the above two equations that

(55)
$$EV^2(T) = v^2Q(t) + E\int_t^T [\pi^*Q\sigma\sigma^*\pi + 2QV\tilde{b}^*\pi + \tilde{b}^*(R + \sigma\sigma^*Q)^{-1}\tilde{b}Q^2V^2]\,ds.$$

Hence,

(56) $J(\pi) = v^2Q(t) + E\int_t^T \tilde{\pi}^*(R + Q\sigma\sigma^*)\tilde{\pi}(s)\,ds - 2E[V(T)Q(T)\xi] + E\xi^2$

with

(57) $$\tilde{\pi}(s) := \pi(s) + (R + Q\sigma\sigma^*)^{-1}\tilde{b}VQ(s), \quad s \in [0,T].$$

Note that

(58)
$$\begin{cases} d[VQ(s)] &= [-rVQ(s) + \tilde{b}^*\pi Q(s) + \tilde{b}^*(R + \sigma\sigma^*Q)^{-1}\tilde{b}QVQ(s)]\,ds \\ &\quad + Q\pi^*\sigma(s)\,dW(s), \quad s \in (t,T], \\ VQ(t) &= vQ(t), \quad VQ(T) = V(T) \end{cases}$$

and equivalently that

(59)
$$\begin{cases} d[VQ(s)] &= [-rVQ(s) + Q\tilde{b}^*\tilde{\pi}(s)]\,ds \\ &\quad + [-QV\tilde{b}^*(R + Q\sigma\sigma^*)^{-1}\sigma Q(s) + Q\tilde{\pi}^*\sigma(s)]\,dW(s), \quad s \in (t,T], \\ VQ(t) &= vQ(t), \quad VQ(T) = V(T) \end{cases}$$

Similarly, we can use the adapted adjoint process (p,q) for the wealth process (51) to express $E[V(T)Q(T)\xi]$ as follows:

(60) $E[V(T)Q(T)\xi] = vp(t) + E\int_t^T Qp\tilde{b}^*\tilde{\pi}(s)\,ds + E\int_t^T Q\sigma q\tilde{\pi}(s)\,ds$

where (p, q) is the unique $\mathcal{F}_t$-adapted solution to the backward stochastic differential equation (47). Then, it follows that

$$
\begin{aligned}
J(\pi) &= [v^2 Q(t) - 2vQp(t)] + E \int_t^T \tilde{\pi}^*(R + Q\sigma\sigma^*)\tilde{\pi}(s)\, ds \\
&\quad -2E \int_t^T Qp\tilde{b}^* \tilde{\pi}(s)\, ds - 2E \int_t^T Q\sigma q \tilde{\pi}(s)\, ds + E\xi^2 \\
&= [v^2 Q(t) - 2vQp(t)] + E \int_t^T \widehat{\pi}^*(R + Q\sigma\sigma^*)\widehat{\pi}(s)\, ds \\
&\quad -E \int_t^T (Qp\tilde{b} + Q\sigma q)^*(R + Q\sigma\sigma^*)^{-1}(Qp\tilde{b} + Q\sigma q)\, ds
\end{aligned}
$$

(61)

with

(62) $$\widehat{\pi}(s) := \tilde{\pi}(s) - (R + Q\sigma\sigma^*)^{-1}(Qp\tilde{b} + Q\sigma q)(s).$$

Since $Q > 0$, the value function has the formula (46), *the hedger's subjective price process* on the contingent claim ξ is

$$\bar{v}(t) = p(t), \quad t \in [0, T],$$

and *the optimal hedging process* has the feedback form (50). The proof is complete.

Remark 1. *Note that for the case discussed in Section 4, the solution (p, q) of the backward stochastic differential equation (47) is characterized by the solution $\bar{v}$ of the partial differential equation (41) in the following way*

(63) $$p(t) = \bar{v}(t, X(t)), \quad q(t) = \bar{v}_x(t, X(t))\sigma^*(t)X(t), \quad 0 \le t \le T.$$

For the details of this connection, the reader is referred to Peng [**22, 23**], Pardoux and Peng [**20**], and Pardoux and Tang [**21**].

6. The Case of Random Market Conditions

In this paper, the market conditions (the instantaneous interest rate, the appreciation rate vector, and the volatility matrix) are assumed to be deterministic. In this case, the pricing formula is associated with a deterministic Riccati equation. When the market conditions are random, the pricing formula is associated with a backward stochastic Riccati equation. The latter more complex case will be exposed elsewhere in details.

References

[1] Barron, E. and Jensen, R., *A stochastic control approach to the pricing of options*, Mathematics of Operations Research, **15** (1990), 49–79

[2] Bertsekas, D. P., *Necessary and sufficient conditions for existence of an optimal portfolio*, J. Economic Theory, **8** (1974), 234–247

[3] Bismut, J. M., *Growth and optimal intertemporal allocation of risks*, J. Economic Theory, **10** (1975), 239–257

[4] Cox, J. and Huang, C. F., *Optimal consumption and portfolio policies when asset prices follow a diffusion process*, J. Economic Theory, **49** (1989), 33–83

[5] Cvitanic, J. and Karatzas, I., *Convex Duality in Constrained Portfolio Optimization*, The Annals of Applied Probability, **2** (1992), 767–818

[6] Cvitanic, J. and Karatzas, I., *Hedging Contingent Claims with Constrained Portfolios*, The Annals of Applied Probability, **2** (1993), 652–681

[7] Duffie, D. and Epstein, L., *Stochastic differential utility*, Econometrica, **60** (1992), 353–394

[8] Duffie, D. and Richardson, H. R., *Mean-variance hedging in continuous time*, Ann. Appl. Probab., **1** (1991), 1–15.

[9] El Karoui, N. and Quenez, M. C., *Dynamic programming and pricing of contingent claims in an incomplete markets*, SIAM J. Control Optim., **33** (1993), 29–66

[10] El Karoui, N., Peng, S. and Quenez, M. C., *Backward stochastic differential equations in finance*, Mathematical Finance, **7** (1997), 1–71.

[11] Föllmer, H. and Schweizer, M., *Hedging of Contingent Claims under Incomplete Information*. In: Applied Stochastic Analysis, Eds. Davis, M. H. A. and Elliott, R. J., Stochastics Monographs **5**, 1991.

[12] Karatzas, I., *Optimization problems in the theory of continuous trading*, SIAM J. Control Optim., **27** (1989), 1221–1259

[13] Karatzas, I., Lehoczky, J. P. and Shreve, S. E., *Optimal portfolio and consumption decisions for a "small investor" on a finite horizon*, SIAM. J. Control Optim., **25** (1987), 1557–1586

[14] Karatzas, I., Lehoczky, J. P., Shreve, S. E. and Xu, G. L., *Martingale and duality methods for utility maximization in an incomplete market*, SIAM J. Control Optim., **29** (1991), 702–730

[15] Kohlmann, M. and Tang, S., *Optimal control of linear stochastic systems with singular costs, and the mean-variance hedging problem with stochastic market conditions*. submitted

[16] Kohlmann, M. and Zhou, X., *Relationship between backward stochastic differential equations and stochastic controls: a linear-quadratic approach*, SIAM J. Control Optim., **38** (2000), 1392–1407

[17] Kwok, Y.-K., Mathematical Models of Financial Derivatives, Springer-Verlag, 1998.

[18] Merton, R. C., Continuous-Time Finance, Basil Black well Inc., 1990.

[19] Pardoux, E. and Peng, S., *Adapted solution of a backward stochastic differential equation*, Systems Control Letters, **14** (1990), 55–61

[20] Pardoux, E. and Peng, S., *Backward stochastic differential equations and quasi-linear parabolic partial differential equations,* in: Rozovskii, B. L., Sowers, R. S. (eds.) Stochastic Partial Differential Equations and Their Applications, Lecture Notes in Control and Information Science 176, Springer, Berlin, Heidelberg, New York 1992, 200–217

[21] Pardoux, E. and Tang, S., *Forward-backward stochastic differential equations with application to quasi-linear partial differential equations of second-order*, Probability Theory and Related Fields, **114** (1999), 123–150

[22] Peng, S., *Probabilistic interpretation for systems of semilinear parabolic PDEs*, Stochastics & Stochastic Reports, **37** (1991), 61–74

[23] Peng, S., *A generalized dynamic programming principle and Hamilton-Jacobi-Bellman equation*, Stochastics & Stochastics Reports, **38** (1992), 119–134

[24] Richardson, H. R., *A minimum variance result in continuous trading portfolio optimization*, Management Sci., **35** (1989), 1045–1055.

[25] Schweizer, M., *Approximating random variables by stochastic integrals*, The Annals of Probability, **22** (1994), 1536–1575

[26] Tang, S., *Deriving Black-Scholes formula from optimality*, preprint, 1998, submitted

[27] Zhou, X. and Li, D., *Explicit efficient frontier of a continuous-time mean-variance portfolio selection problem,* In: Contrtol of Distributed and Stochastic Systems (eds: S. Chen, X. Li, J. Yong and X. Zhou), Kluwer Academic Publishers, 1999.

DEPARTMENT OF MATHEMATICS, AND THE LABORATORY OF MATHEMATICS FOR NONLINEAR SCIENCES AT FUDAN UNIVERSITY, FUDAN UNIVERSITY, SHANGHAI 200433, CHINA.
E-mail address: sjtangk@online.sh.cn, tang@fmi.uni-konstanz.de

AMS/IP Studies in Advanced Mathematics
Volume 26, 2002

An Overview on the Martingale Approach to Option Pricing

Jia-an Yan

ABSTRACT. The option pricing theory has its origin in the seminal papers by Black and Scholes (1973) and by Merton (1973). Harrison and Kreps (1979) and Harrison and Pliska (1981) have showed that a natural mathematical framework for the analysis of financial markets is martingale theory and stochastic analysis. Since then this framework has played a dominating role in option pricing. In this note, we give an overview on the martingale approach to option pricing in the Black-Scholes model as well as in the semimartingale model. A new look at the fundamental theorem of asset pricing is given. Several methods of choosing the martingale measure for option pricing in incomplete markets are presented.

0. Introduction

The option pricing has its origin in the seminal papers by Black and Scholes (1973) and by Merton (1973), where Itô's formula has been used for deriving the Black-Scholes equation and the risk neutral valuation principle was introduced. Harrison and Kreps (1979) and Harrison and Pliska (1981) have further showed that a natural mathematical framework for the analysis of financial markets is martingale theory and stochastic analysis. Since then this framework has played a dominating role in option pricing. This theory has become a powerful and effective tool for quantitative analysis in finance and in many economical problems.

An option or contingent claim is a financial contract written on an underlying asset (such as stock, foreign currency, stock-index, or derivative security). In option pricing theory, a basic assumption is that the market is *arbitrage free*, meaning that there is no riskless profitable opportunity in the market. In such a market, it is reasonable to price a contingent claim in such a manner that the market augmented with the resulting derivative security remains arbitrage free. Such a pricing principle is called *pricing by arbitrage* or *arbitrage pricing*. It turns out that the martingale theory and stochastic analysis is a proper framework for characterizing arbitrage-free market and for pricing contingent claims.

In this note, we will give an overview on the martingale approach to option pricing. For clarity and simplicity, we only consider European contingent claims in a *frictionless* market. By latter we mean there are no transaction costs, no

Work supported in part by City University of Hong Kong, contract 9000906, and by National Science Foundation of China, grant 79790130.

bid-ask spread, no restrictions on trade such as margin requirements or short sale restrictions, there are no taxes, borrowing and lending are at the same interest rate, and asset shares are divisible. In addition, we assume that trading in assets takes place continually in time.

The following two theorems will be used in the sequel.

Theorem 0.1 Let $\mathcal{G}$ be a sub-σ-algebra of $\mathcal{F}$, $g(x,y)$ a non-negative Borel function on $\mathbf{R}^m \times \mathbf{R}^n$. If X is a $\mathcal{G}$-measurable $\mathbf{R}^m$-valued r.v. and Y is an $\mathbf{R}^n$-valued r.v. independent of $\mathcal{G}$, then

$$\mathbf{E}[g(X,Y) \mid \mathcal{G}] = \mathbf{E}[g(x,Y)]|_{x=X}. \tag{0.1}$$

Theorem 0.2 (*Bayes' rule*)   Let $\mathbf{Q}$ be a probability measure equivalent to $\mathbf{P}$ and $\mathcal{G}$ a sub-σ-algebra of $\mathcal{F}$. We put $\xi = \frac{d\mathbf{Q}}{d\mathbf{P}}$, $\eta = \mathbf{E}[\xi \mid \mathcal{G}]$. Then for a $\mathbf{Q}$-integrable r.v. X we have

$$\mathbf{E_Q}[X \mid \mathcal{G}] = \eta^{-1}\mathbf{E}[X\xi \mid \mathcal{G}]. \tag{0.2}$$

The rest of this note is organized as follows: In Section 1 we introduce the Black-Scholes economy and present the martingale approach to option pricing with an illustrative example. In Sections 2 we introduce the semimartingale model and give a new look at the fundamental theorem of asset pricing. An Itô's model is presented in Section 3. In Section 4 we present the main results on the martingale approach to option pricing. As an illustrative example, the pricing of foreign currency option is given. Finally, in Section 5 we present several methods of choosing the martingale measure for option pricing in incomplete markets.

1. Black-Scholes Economy

1.1 The Black-Scholes Model

We consider the Black-Scholes economy in which there are two financial instruments: a risky asset (stock), and a riskless asset (savings account). Assume that the stock pays no dividends and its price process satisfies the Itô SDE

$$dS_t = S_t(\mu dt + \sigma dB_t), \tag{1.1}$$

where $S_0 > 0$, μ and σ are constants, (B_t) is a Brownian motion defined on a filtered probability space $(\Omega, \mathcal{F}, (\mathcal{F}_t), \mathbf{P})$, where $\mathcal{F}_t = \sigma(B_s, s \leq t)$. Such a process (S_t) is called a *geometric Brownian motion*. It is also called a *log-normal* process because

$$S_t = S_0 \exp\left\{(\mu - \frac{\sigma^2}{2})t + \sigma B_t\right\}, \tag{1.2}$$

so that $\log(S_t)$ is normally distributed. We call μ the (*instantaneous*) *expected rate of return* and σ the *volatility* of S. The value process (β_t) of the savings account is assumed to satisfy

$$d\beta_t = r\beta_t dt, \tag{1.3}$$

where r is the constant interest rate. In the sequel, we always assume $\beta_0 = 1$ so that $\beta_t = e^{rt}$.

A *trading strategy* is a pair of $\mathcal{F}_t$-adapted processes $\{a, b\}$, where $a(t)$ denotes the number of units of the asset held at time t, $b(t)\beta_t$ is the total money invested in the bank at time t. So the *wealth* V_t at time t of a portfolio $\{a(t), b(t)\}$ is given by

$$V_t = a(t)S_t + b(t)\beta_t.$$

A trading strategy $\{a, b\}$ is said to be *self-financing* if the change of its wealth is only due to the changes in the asset prices, meaning that

$$dV_t = a(t)dS_t + b(t)d\beta_t. \tag{1.4}$$

A trading strategy is said to be *admissible*, if its wealth process (V_t) is non-negative.

1.2 Martingale Measure and Pricing by Arbitrage

As stated in the introduction, in option pricing theory a basic assumption on the market is the absence of arbitrage opportunity. We will show that the Black-Scholes economy is arbitrage free. More precisely, we will show that there is no admissible self-financing strategy which has initial wealth zero and a non-negative terminal wealth V_T with $\mathbf{P}(V_T > 0) > 0$.

Lemma 1.1 A trading strategy $\{a, b\}$ is self-financing if and only if its discounted wealth process $(\widetilde{V}_t)$ satisfies

$$d\widetilde{V}_t = a(t)d\widetilde{S}_t. \tag{1.5}$$

Proof Assume $\{a, b\}$ to be self-financing. Since $\widetilde{V}_t = e^{-rt}V_t$, by (1.3) we have

$$
\begin{aligned}
d\widetilde{V}_t &= -V_t r e^{-rt} dt + e^{-rt} dV_t \\
&= -[a(t)S_t + b(t)e^{rt}]re^{-rt}dt + e^{-rt}[a(t)dS_t + b(t)e^{rt}rdt] \\
&= a(t)[S_t d(e^{-rt}) + e^{-rt}dS_t] = a(t)d\widetilde{S}_t.
\end{aligned}
$$

Similarly, we can prove "if" part.

Now we show that there exists a probability measure $\mathbf{P}^*$ equivalent to $\mathbf{P}$ such that the process $\widetilde{S}_t$ is a $\mathbf{P}^*$-martingale. In fact, we can rewrite (1.1) as

$$d\widetilde{S}_t = \widetilde{S}_t[(\mu - r)dt + \sigma dB_t].$$

Consequently, if we put $\frac{d\mathbf{P}^*}{d\mathbf{P}}|_{\mathcal{F}_T} = \exp\left\{ -\frac{\mu-r}{\sigma}B_T - \frac{1}{2}\left(\frac{\mu-r}{\sigma}\right)^2 T\right\}$, then by the Girsanov's theorem $B_t^* = B_t + \frac{\mu-r}{\sigma}t$ is a $\mathbf{P}^*$-Brownian motion and

$$d\widetilde{S}_t = \widetilde{S}_t \sigma dB_t^*. \tag{1.6}$$

Thus $(\widetilde{S}_t)$ is a $\mathbf{P}^*$-martingale. We call this probability measure $\mathbf{P}^*$ a *martingale measure* for the market. By contrast, $\mathbf{P}$ is called the *objective* or *physical* probability measure. It is easy to see that the martingale measure for the Black-Scholes economy is unique.

On the other hand, (1.1) can be rewritten as

$$dS_t = S_t[rdt + \sigma dB_t^*]. \tag{1.7}$$

It means that under this measure $\mathbf{P}^*$ the expected rate of return of the risky asset is equal to the interest rate of the savings account. For this reason the martingale measure $\mathbf{P}^*$ is also called a *risk-neutral probability measure*.

By Lemma 1.1, the discounted wealth process of a self-financing strategy is a local martingale under $\mathbf{P}^*$. Thus, for any admissible self-financing strategy, its discounted wealth process is a $\mathbf{P}^*$-supermartingale, because it is a non-negative local $\mathbf{P}^*$-martingale. Consequently, if the initial wealth of an admissible self-financing strategy is zero then its wealth at any time must be zero. This implies that the market has no arbitrage.

The following theorem is the main result of the arbitrage pricing in the Black-Scholes economy.

Theorem 1.2 Let ξ be a European contingent claim which is integrable under $\mathbf{P}^*$. Then there exists an admissible self-financing strategy $\{a, b\}$ replicating ξ such that its wealth process (V_t) is given by

$$V_t = \mathbf{E}^*\big[e^{-r(T-t)}\xi|\mathcal{F}_t\big], \tag{1.8}$$

or equivalently, $(\widetilde{V}_t)$ is a $\mathbf{P}^*$-martingale. If $V_t = F(t, S_t)$ with $F \in C^{1,2}([0, T) \times \mathbf{R}_+$, then

$$a(t) = F_x(t, S_t). \tag{1.9}$$

Proof We define V_t by (1.8). Then $(\widetilde{V}_t)$ is a $\mathbf{P}^*$- martingale. Since $(\mathcal{F}_t)$ is also the natural filtration of (B_t^*), by the martingale representation theorem for Brownian motion there exists an $H \in \mathcal{L}^2$ such that

$$\widetilde{V}_t = V_0 + \int_0^t H_s dB_s^*, \quad t \in [0, T]. \tag{1.10}$$

Put

$$a(t) = H_t/(\sigma \widetilde{S}_t), \quad b(t) = \widetilde{V}_t - a(t)\widetilde{S}_t. \tag{1.11}$$

Then $\{a, b\}$ is a replicating strategy for ξ and (V_t) is its wealth process. By (1.6) and (1.11) we have

$$a(t)d\widetilde{S}_t = a(t)\widetilde{S}_t \sigma dB_t^* = H_t dB_t^* = d\widetilde{V}_t.$$

Thus by Lemma 1.1 the strategy $\{a, b\}$ is self-financing and admissible.

Now assume that $V_t = F(t, S_t)$, where $F \in C^{1,2}([0, T) \times \mathbf{R}^+$. By Itô's formula,

$$
\begin{aligned}
d\widetilde{V}_t &= d\Big(e^{-rt}F(t, S_t)\Big) = e^{-rt}F_x(t, S_t)dS_t + \text{``dt'' term} \\
&= F_x(t, S_t)d\widetilde{S}_t + \text{``dt'' term}.
\end{aligned}
$$

Since $(\widetilde{V}_t)$ and $\widetilde{S}_t$ are $\mathbf{P}^*$-martingales, the "dt" term must vanish. Thus by Lemma 1.1 we know (1.9) holds.

Remark It is natural to define V_t as the *fair price* at time t of the contingent claim ξ, because with this price there does not exist any arbitrage opportunity for both the seller and buyer of the contingent claim. Equation (1.8) is called the *risk-neutral valuation* formula.

Corollary 1.3 If $\xi = f(S_T)$, then $V_t = F(t, S_t)$, where

$$F(t, x) = e^{-r(T-t)} \int_{-\infty}^{\infty} f\Big(xe^{(r-\sigma^2/2)(T-t)+\sigma y\sqrt{T-t}}\Big)\frac{e^{-y^2/2}}{\sqrt{2\pi}}dy. \tag{1.12}$$

Proof We express S_T as

$$S_T = S_t(S_T S_t^{-1}) = S_t \exp\{(r - \sigma^2/2)(T - t) + \sigma(B_T^* - B_t^*)\}.$$

Since S_t is measurable w.r.t. $\mathcal{F}_t$ and $B_T^* - B_t^*$ is independent of $\mathcal{F}_t$, by (1.8) and Theorem 0.1 we have

$$V_t = \mathbf{E}^*\Big[e^{-r(T-t)}f(x\exp\{(r - \sigma^2/2)(T - t) + \sigma(B_T^* - B_t^*)\})\Big]\Big|_{x=S_t},$$

from which we get $V_t = F(t, S_t)$.

Now we consider a European call option $\xi = (S_T - K)^+$, whose price at time t is denoted by $V_t = C(t, S_t)$. From (1.12) we get immediately the following Black-Scholes formula :

$$C(t, x) = xN(d_1) - Ke^{-r(T-t)}N(d_2), \tag{1.13}$$

where $N(z)$ is the standard normal cumulative distribution function and

$$d_1 = \frac{\log(x/K) + (r + \frac{1}{2}\sigma^2)(T - t)}{\sigma\sqrt{T - t}}, \quad d_2 = \frac{\log(x/K) + (r - \frac{1}{2}\sigma^2)(T - t)}{\sigma\sqrt{T - t}}.$$

1.4   An illustative example

We illustrate a technical point for the martingale approach to option pricing through lookback options: find a suitable multiplicative decomposition of the contingent claim so that one can use Theorem 0.1 to compute the conditional expectation. For reader's convenience we give the details of derivations (see Yan (1998a)).

By a lookback option we mean an option whose payoff depends on the maximum or minimum of the realized asset price over the option's life. There are two types of options: the *lookback strike* and *lookback rate*, for both call and put. We only consider the lookback strike put option, whose payoff is defined by

$$\eta = \max_{0 \leq s \leq T} S_s - S_T.$$

By (1.8) the price at time t of the option η is given by

$$P_t = e^{-r(T-t)}\mathbf{E}^*[\eta \mid \mathcal{F}_t].$$

Put

$$M_t = \max_{0 \leq s \leq t} S_s, \quad L_t = \max_{t \leq s \leq T} S_s.$$

Then M_t is $\mathcal{F}_t$-measurable. Let $\lambda = r - \frac{1}{2}\sigma^2$. Since

$$S_t^{-1} L_t = \exp\{\max_{t \leq s \leq T}(\sigma(B_s^* - B_t^*) + \lambda(s - t))\},$$

$S_t^{-1} L_t$ is independent of $\mathcal{F}_t$. Using these notations and by Theorem 0.1 we have

$$\begin{aligned}
P_t &= e^{-r(T-t)}\mathbf{E}^*[M_T - S_T|\mathcal{F}_t] = e^{-r(T-t)}S_t\mathbf{E}^*[\max(S_t^{-1}M_t, S_t^{-1}L_t)|\mathcal{F}_t] - S_t \\
&= e^{-r(T-t)}S_t\mathbf{E}^*[\max(y, S_t^{-1}L_t)]|_{y=S_t^{-1}M_t} - S_t .
\end{aligned}$$

We denote $\mathbf{P}\left(\max_{s \leq t}(\sigma B_s + \lambda s) \leq x\right)$ by $F_t(x)$, then

$$F_t(x) = N\left(\frac{x - \lambda t}{\sigma\sqrt{t}}\right) - e^{2\lambda x/\sigma^2} N\left(\frac{-x - \lambda t}{\sigma\sqrt{t}}\right),$$

where $x \geq 0$ and $N(z)$ is the cumulative standard normal distribution. Thus we have

$$\begin{aligned}
\mathbf{E}^*[\max(y, S_t^{-1}L_t)] &= \mathbf{E}^*\left[\exp\{\max(\log y, \max_{t \leq s \leq T}[\sigma(B_s^* - B_t^*) + \lambda(s - t)])\}\right] \\
&= \mathbf{E}\left[\exp\{\max(\log y, \max_{0 \leq s \leq T-t}(\sigma B_s + \lambda s))\}\right] \\
&= yF_{T-t}(\log y) + \int_{\log y}^{\infty} e^x F'_{T-t}(x)dx.
\end{aligned}$$

Finally we get the following formula, due to Goldman-Sosin-Gatto (1979):

$$\begin{aligned}
P_t = \; &S_t(-1 + N(d_3)(1 + \sigma^2/2r)) \\
&+ M_t e^{-r(T-t)}\left(N(d_1) - \frac{\sigma^2}{2r}(S_t^{-1}M_t)^{(2r/\sigma^2)-1}N(d_2)\right),
\end{aligned}$$

where

$$d_1 = \frac{\log(M_t/S_t) - (r - \frac{1}{2}\sigma^2)(T-t)}{\sigma\sqrt{T-t}},$$

$$d_2 = \frac{-\log(M_t/S_t) - (r - \frac{1}{2}\sigma^2)(T-t)}{\sigma\sqrt{T-t}},$$

$$d_3 = \frac{-\log(M_t/S_t) + (r + \frac{1}{2}\sigma^2)(T-t)}{\sigma\sqrt{T-t}}.$$

2. Semimartingale Model

We fix a finite time-horizon $[0,T]$ and consider a security market which consists of $m+1$ assets whose price processes $(S_t^i), i = 0, \cdots, m$ are assumed to be strictly positive semimartingales, defined on a filtered probability space $(\Omega, \mathcal{F}, \mathcal{F}_t, \mathbf{P})$ satisfying the usual conditions. Moreover, we assume that $\mathcal{F}_0$ is the trivial σ-algebra. In option pricing one needs to choose one asset as a common unit, on the basis of which the prices of other assets are expressed. The resulting relative prices are called the *deflated prices* or *denominated prices*. Such an asset is called a *numeraire asset* or *numeraire*, and its price process is called a *numeraire*. If one takes the savings account as a numeraire asset, the relative prices are usually called *discounted prices*.

For notational convenience, we take asset 0 as the numeraire asset. We set $\gamma_t \hat{=} (S_t^0)^{-1}$ and call γ_t the *deflator* at time t. We set $S_t = (S_t^1, \cdots, S_t^m)$ and $\widetilde{S}_t = (\widetilde{S}_t^1, \cdots, \widetilde{S}_t^m)$, where $\widetilde{S}_t^i = \gamma_t S_t^i, 1 \leq i \leq m$. We call $(\widetilde{S}_t)$ the *deflated* price process of the assets. Note that the deflated price process of asset 0 is the constant 1.

In order to be able to define a trading strategy we need the notion of integration w.r.t. a vector-valued semimartingale. Such integral is defined globally and not componentwise. If $H = (H^0, \cdots, H^m)$ is integrable w.r.t. a semimartingale $(X^0, \cdots, X^m)$ and H^0 is integrable w.r.t. X^0, then we have

$$H.(X^0, \cdots, X^m) = H^0.X^0 + (H^1, \cdots, H^m).(X^1, \cdots, X^m). \tag{2.1}$$

A *trading strategy* is a $\mathbf{R}^{m+1}$-valued $\mathcal{F}_t$-predictable process $\phi = \{\theta^0, \theta\}$, where

$$\theta(t) = (\theta^1(t), \cdots, \theta^m(t)),$$

such that ϕ is integrable w.r.t semimartingale (S^0, S) with $S = (S^1, \cdots, S^m)$. $\theta^i(t)$ represents the numbers of units of asset i held at time t. The wealth $V_t(\phi)$ at time t of a trading strategy $\phi = \{\theta^0, \theta\}$ is

$$V_t(\phi) = \theta^0(t)S_t^0 + \theta(t) \cdot S_t, \tag{2.2}$$

where $\theta(t) \cdot S_t = \sum_{i=1}^m \theta^i(t)S_t^i$. The deflated wealth at time t is $\widetilde{V}_t(\phi) = V_t(\phi)\gamma_t$. A trading strategy $\{\theta^0, \theta\}$ is said to be *self-financing*, if

$$V_t(\phi) = V_0(\phi) + \int_0^t \phi(u)d(S_u^0, S_u). \tag{2.3}$$

We use notation $(H.X)_t$ to stand for the integral of H w.r.t. X over the interval $(0,t]$. In particular, we have $(H.X)_0 = 0$.

Definition 2.1 A security market is said to be *fair* if there exists a probability measure $\mathbf{Q}$ equivalent to the "objective" probability measure $\mathbf{P}$ such that the deflated price processes $(\widetilde{S}_t)$ is a (vector-valued) $\mathbf{Q}$-martingale. We call such a $\mathbf{Q}$ an *equivalent martingale measure* for the market.

We denote by $\mathcal{M}^j$ the set of all equivalent martingale measures for the market, if (S_t^j) is taken as a numeraire.

Assume that $\mathcal{M}^0 \neq \emptyset$. Let (X_t) be the wealth process of an admissible self-financing strategy such that (X_t) is strictly positive and $(\widetilde{X}_t)$ is a $\mathbf{P}^*$-martingale for some $\mathbf{P}^* \in \mathcal{M}^0$. We denote by $\mathcal{M}^X$ the set of all equivalent martingale measures for the market, if (X_t) is taken as a numeraire. We define a probability measure $\mathbf{Q} := h(\mathbf{P}^*)$ by

$$\frac{d\mathbf{Q}}{d\mathbf{P}^*} = \frac{S_0^0}{X_0}(S_T^0)^{-1}X_T. \tag{2.4}$$

By Bayes' rule it is easy to show that h is a bijection from $\mathcal{M}^0$ onto $\mathcal{M}^X$. In particular, this result implies the following

Theorem 2.2 The fairness of a market is independent of the choice of numeraire.

A strategy is said to be *admissible*, if its wealth process is non-negative. A strategy is said to be *tame*, if its deflated wealth process is bounded from below by some real constant. The weakness of the notion of tame strategy is that it is not invariant under the change of numeraire. Moreover, all bounded elementary or simple strategies are not tame. For this reason we have proposed in Yan (1998b) to extend the notion of tame strategy to a notion of "allowable strategy".

Definition 2.3 A strategy $\phi = \{\theta^0, \theta\}$ is said to be *allowable*, if there exists a positive constant c such that the wealth $(V_t(\phi))$ at any time t is bounded from below by $-c\sum_{i=0}^m S_t^i$.

Definition 2.4 A market is said to have no arbitrage with allowable strategies if there exists no allowable self-financing strategy with initial wealth zero and a non-negative terminal wealth V_T such that $\mathbf{P}(V_T > 0) > 0$.

A key point of arbitrage pricing of contingent claims is the following characterization of the self-financing strategy (see Yan (1998b) for a proof).

Theorem 2.5 A strategy $\phi = \{\theta^0, \theta\}$ is self-financing if and only if its wealth process (V_t) satisfies

$$d\widetilde{V}_t = \theta(t)d\widetilde{S}_t, \tag{2.5}$$

where $\widetilde{V}_t = V_t\gamma_t$. In particular, the deflated wealth process of an allowable self-financing strategy is a local $\mathbf{Q}$-martingale and a $\mathbf{Q}$-supermartingale for any $\mathbf{Q} \in \mathcal{M}^0$.

As a corollary we obtain

Theorem 2.6 A fair market has no arbitrage with allowable strategies.

Proof Let $\mathbf{Q} \in \mathcal{M}^0$. Let $\{\theta, \theta^0\}$ be an allowable self-financing strategy with initial wealth zero. By Theorem 2.5 the deflated wealth process of ϕ is a $\mathbf{Q}$-supermartingale. Therefore, we must have $\mathbf{E}_\mathbf{Q}[\widetilde{V}_T] \leq 0$. So the market has no arbitrage with allowable strategies.

From a general version of the fundamental theorem of asset pricing, due to Delbaen and Schachermayer (1994), we can deduce the following characterization of the fairness of a market (see Yan (1998b)). This characterization is intrinsic in the sense that it does not depend on the choice of numeraires.

Theorem 2.7 The market is fair if and only if there is no sequence (ϕ_n) of allowable self-financing strategies with initial wealth 0 such that $V_T(\phi_n) \geq$

$-\frac{1}{n}\sum_{i=0}^{m} S_T^i$ a.s., for all $n \geq 1$ and such that $V_T(\phi_n)$ a.s. tends to a non-negative random variable ξ satisfying $\mathbf{P}(\xi > 0) > 0$.

Remark In Delbaen and Schachermayer (1994) a similar condition, called the *no free lunch with vanishing risk* condition, has been formulated in terms of tame strategies and deflated price process, and is shown to be equivalent to the existence of a "local martingale measure". Since then many authors adopt this framework. But there are two shortcomings: the one is this framework is not invariant under the change of numeraires; another one is in such a market it may have arbitrage opportunity for allowable strategies. For example, consider a Merton's model, in which the stock price process satisfies the Itô SDE

$$dS_t = S_t(rdt + \sigma_t dB_t),$$

where σ_t is a strictly positive adapted process. If $\widetilde{S}_t$ is only a local martingale but not a martingale, then $S_0 > \mathbf{E}[\widetilde{S}_T]\hat{=}x$. With initial wealth S_0 one can replicate $\frac{S_0}{x}S_T$ at time T. So a short salling of the stock leads to an arbitrage strategy.

In general, for a fair market the martingale measures are not unique. If for a given numeraire the martingale measure $\mathbf{P}^*$ is unique, then according to the martingale representation theorem in stochastic analysis, each European contingent claim having a $\mathbf{P}^*$-integarable deflated value, can be replicated by an admissible self-financing strategy. A market having this property is called a *complete market*. In principle, a fair market is complete if and only if the martingale measure is unique (see Stricker and Yan (1998) for a general result in this direction).

3. The Itô Process Model

Now we consider a concrete case of semimartingale model: the Itô process model. This model is widely adopted in financial economy. We fix a finite time-horizon T. Let $B = (B^1, \cdots, B^d)$ be a Brownian motion on a complete probability space $(\Omega, \mathcal{F}, \mathbf{P})$. We denote by $(\mathcal{F}_t)$ the natural filtration of (B_t) and by $\mathcal{L}$ the set of all measurable $(\mathcal{F}_t)$-adapted processes. We consider a financial market which consists of $m + 1$ assets. The price process (S_t^i) of each asset i is assumed to be a strictly positive Itô process. Since its logarithm is also an Itô process, we can represent (S_t^i) as

$$dS_t^i = S_t^i\Big[\sigma^i(t)dB_t + \mu^i(t)dt\Big], \quad S_0^i = p_i, \quad 0 \leq i \leq m. \tag{3.1}$$

We call $\mu = (\mu^0, \cdots, \mu^m)$ the *vector of expected rate of return* and σ the *volatility matrix*.

We specify asset 0 as the numeraire asset and set $\gamma_t := (S_t^0)^{-1}$. By Itô's formula we have

$$d\gamma_t = -\gamma_t\Big[\sigma^0(t)dB_t + (\mu^0(t) - |\sigma^0(t)|^2)dt\Big], \tag{3.2}$$

$$d\widetilde{S}_t^i = \widetilde{S}_t^i\Big[a^i(t)dB_t + b^i(t)dt\Big], \quad 1 \leq i \leq m, \tag{3.3}$$

where

$$a^i(t) = \sigma^i(t) - \sigma^0(t); \quad b^i(t) = \mu^i(t) - \mu^0(t) + |\sigma^0(t)|^2 - \sigma^i(t) \cdot \sigma^0(t).$$

In particular, if asset 0 is a savings account with interest rate process $(r(t))$, then

$$a^i(t) = \sigma^i(t), \quad b^i(t) = \mu^i(t) - r(t).$$

One raises naturally a question: what conditions we should impose on the coefficients a and b of the diffusion $(\widetilde{S}_t)$ such that the market is fair? The following theorem gives a partial answer to this question (see Karatzas (1996) or Yan (1998b) for a proof).

Theorem 3.1 If the market is fair, the linear equation

$$a(t)\psi(t) = b(t), \quad dt \times d\mathbf{P}-\text{a.e.}, \text{a.s. on } [0, T] \times \Omega \tag{3.4}$$

has a solution $\psi \in (\mathcal{L}^2)^d$, where $\mathcal{L}^2$ stands for the set of all adapted process ϕ with $\int_0^T \phi^2(u)du < \infty$. Conversely, if

$$\mathbf{E}\left[\exp\left\{\frac{1}{2}\int_0^T |a^i(t)|^2 dt\right\}\right] < \infty, \quad 1 \le i \le m, \tag{3.5}$$

and equation (3.4) has a solution $\psi \in (\mathcal{L}^2)^d$ satisfying

$$\mathbf{E}\left[\exp\left\{\frac{1}{2}\int_0^T |\psi(t)|^2 dt\right\}\right] < \infty, \tag{3.6}$$

then the probability measure $\mathbf{Q}$ with Radon-Nikodym derivative $\frac{d\mathbf{Q}}{d\mathbf{P}} = \mathcal{E}(-\psi.B)_T$ is an equivalent martingale measure.

Remark Without condition (3.5) $\mathbf{Q}$ is an equivalent local martingale measure but not necessarily a martingale measure. We refer the reader to Ansel and Stricker (1993) for an investigation on this subject.

The following theorem provides a sufficient condition for the existence of a unique equivalent martingale measures.

Theorem 3.2 Assume that $m \ge d$, a satisfies (3.5) and $a^T(t)a(t)$ are non-degenerated for a.e. t, where $a^T(t)$ stands for the transpose of $a(t)$. Put $\psi(t) = (a^T(t)a(t))^{-1}a^T(t)b(t)$. If ψ satisfies (3.4) and (3.6), then there exists a unique equivalent martingale measure $\mathbf{P}^*$ for the market. Moreover, we have

$$\mathbf{E}\left[\frac{d\mathbf{P}^*}{d\mathbf{P}} \,\Big|\, \mathcal{F}_t\right] = \exp\left\{-\int_0^t \psi(s)dB_s - \frac{1}{2}\int_0^t |\psi(s)|^2 ds\right\}, \; 0 \le t \le T.$$

Proof By Theorem 3.1 there exists an equivalent martingale measure. To prove the uniqueness, let $\mathbf{Q}$ be an equivalent martingale measure. There exists a $\theta \in (\mathcal{L}^2)^d$ such that $\frac{d\mathbf{Q}}{d\mathbf{P}} = \mathcal{E}(-\theta.B)_T$. By Theorem 3.1, we have $a(t)\theta(t) = b(t)$. Consequently, applying $(a^T(t)a(t))^{-1}a^T(t)$ to the both sides of this equation we get $\theta(t) = \psi(t)$. The uniqueness is thus proved.

Remark If $m = d$, then ψ satisfies (3.4) automatically. In this case if (3.5)-(3.6) are satisfied, then there exists a unique equivalent martingale measure if and only if $a(t, \omega)$ is non-singular, for $(t, \omega) \in [0, T] \times \Omega$, a.e., a.s.. See Karatzas (1996).

4. Martingale Approach to Option Pricing

In this section we will present the martingale approach to pricing European contingent claims in a fair market with semimartingale model.

Let ξ be a contingent claim. Assume that $\gamma_T \xi$ is $\mathbf{P}^*$-integrable for some $\mathbf{P}^* \in \mathcal{M}^0$. We put

$$V_t = \gamma_t^{-1}\mathbf{E}^*[\gamma_T \xi \,|\, \mathcal{F}_t]. \tag{4.1}$$

If we consider (V_t) as the price process of an asset, then the market augmented with this asset is still fair, because the deflated price process of this asset is a $\mathbf{P}^*$-martingale. So it seems that (V_t) can be considered as a candidate for a "fair" price process of ξ. This definition depends on the choice of martingale measure. We will show that for replicatable contingent claims the fair price is unique.

Definition 4.1 A European contingent claim ξ is said to be *attainable* if there exists a $\mathbf{P}^* \in \mathcal{M}^0$, such that $\gamma_T \xi$ is $\mathbf{P}^*$-integrable and there exists an admissible self-financing strategy ϕ such that its terminal wealth is equal to ξ and its deflated wealth process is a $\mathbf{P}^*$-martingale (i.e. $\mathbf{E}^*[\gamma_T \xi] = \gamma_0 V_0(\phi)$). In this case, we say that ξ is $\mathbf{P}^*$-*attainable* and call ϕ a $\mathbf{P}^*$-*replicating strategy* for ξ.

The following theorem shows that the "fair" price process of an attainable contingent claim is uniquely determined (see Yan (1998b)).

Theorem 4.2 Let ξ be an attainable contingent claim, and $\mathbf{P}^*, \mathbf{P}' \in \mathcal{M}^0$ be such that ξ is $\mathbf{P}^*$- and $\mathbf{P}'$-attainable. Let (V_t) (resp. (U_t)) be the wealth process of a $\mathbf{P}^*$- (resp. $\mathbf{P}'$-)replicating strategy for ξ. Then (V_t) and (U_t) are the same. Moreover, V_t is given by (4.1).

According to Theorem 4.2, for a $\mathbf{P}^*$-attainable contingent claim ξ it is natural to define its *"fair" price* at time t by (4.1). We call this method of pricing the *arbitrage pricing* (or *pricing by arbitrage*). The following theorem shows that the attainability of a contingent claim and the arbitrage pricing of attainable contingent claims are independent of the choice of numeraire (see Yan (1998b)).

Theorem 4.3 Let $\mathbf{P}^* \in \mathcal{M}^0$ and ξ be a $\mathbf{P}^*$-attainable contingent claim and ϕ be a $\mathbf{P}^*$-replicating strategy for ξ. Then for any $0 \leq j \leq m$ ξ is an $h_j(\mathbf{P}^*)$-attainable contingent claim, and its "fair" price process remains the same.

Remark More generally, assume that (X_t) is a strictly positive wealth process of some self-financing strategy such that $(\widetilde{X}_t)$ is a $\mathbf{P}^*$-martingale. Then ξ is a $\mathbf{Q}$-attainable contingent claim, where $\mathbf{Q}$ is given by (2.4), and its "fair" price process defined by

$$V_t = X_t \mathbf{E}_{\mathbf{Q}}[X_T^{-1} \xi \mid \mathcal{F}_t]$$

coincides with that given by (4.1).

Now we illustrate the martingale method by an example: pricing a foreign currency option.

Consider a contract which give its owner the right to buy M units of a foreign currency at a prespecified exchange rate K and at the date T. We assume that both domestic and foreign risk-free interest rates are non-negative constants, denoted by r^d and r^f, and the exchange rate Q satisfies the following equation:

$$dQ_t = Q_t[\mu dt + \sigma dB_t], \tag{4.2}$$

where μ and σ are constants. From a domestic point of view, there are two basic assets: one is the domestic savings account (regarded as a risk-free asset), whose price process is

$$d\beta_t^d = r^d \beta_t^d dt;$$

another is the foreign savings account in domestic term (regarded as a risky asset), whose price process is $S_t := \beta_t^f Q_t$, where

$$d\beta_t^f = r^f \beta_t^f dt.$$

By Itô's formula, we get

$$dS_t = S_t[(r^f + \mu)dt + \sigma dB_t].$$

Now the contract can be regarded as a European call option in this domestic market, whose payoff at expiry date T is $\xi = M(Q_T - K)^+$.

Following Karatzas (1996) we will derive a formula for the price of the option ξ. We take β_t^d as a numeraire, and let $\widetilde{S}_t = (\beta_t^d)^{-1} S_t$, then

$$d\widetilde{S}_t = \widetilde{S}_t[(r^f - r^d + \mu)dt + \sigma dB_t].$$

Consequently, if we put $\frac{d\mathbf{P}^*}{d\mathbf{P}}|_{\mathcal{F}_T} = \mathcal{E}(-\theta.B)_T$ with $\theta(t) = \theta = (r^f - r^d + \mu)/\sigma$, then by Girsanov's theorem $B_t^* = B_t + \theta t$ is a $\mathbf{P}^*$-Brownian motion and $(\widetilde{S}_t)$ is a $\mathbf{P}^*$-martingale. We call $\mathbf{P}^*$ the *domestic martingale measure*. By (1.8) the price at time t of the option ξ is given by

$$V_t = \mathbf{E}^*[e^{-r^d(T-t)}M(Q_T - K)^+|\mathcal{F}_t]. \tag{4.3}$$

We can rewrite (4.2) and (4.3) as

$$dQ_t = Q_t[(r^d - r^f)dt + \sigma dB_t^*],$$

$$V_t = Me^{-r^f(T-t)}\mathbf{E}^*[e^{-(r^d-r^f)(T-t)}(Q_T - K)^+|\mathcal{F}_t].$$

Thus by the Black-Scholes formula we obtain immediately the valuation formula for option ξ:

$$V_t = Me^{-r^f(T-t)}C(t, Q_t),$$

where $C(t, x)$ is given by (1.13) with r being replaced by $r^d - r^f$. This formula is due to Garman-Kohlhagen (1983).

5. Option Pricing in Incomplete Markets

If the market is incomplete, we can not uniquely determine the price of a non-attainable contingent claim, because martingale measures are not unique. In this case we have the following result, essentially due to Jacka (1992) (see also Stricker and Yan (1998)).

Theorem 5.1 Let ξ be a contingent claim. Put

$$V_0^b = \inf_{Q \in \mathcal{M}^0} E_Q[\gamma_T \xi], \quad V_0^a = \sup_{Q \in \mathcal{M}^0} E_Q[\gamma_T \xi].$$

Then each $x \in (V_0^b, V_0^a)$ is an arbitrage-free price. ξ is attainable if and only if $V_0^b = V_0^a$.

To get round the non-uniqueness of arbitrage-free price several methods of choosing the martingale measure for option pricing have been proposed. We present below four methods.

(1) Numeraire portfolio method   The numeraire portfolio method was initiated by J. Long (1990) and developed by Bajeux and Portait (1995a, 1995b). The starting point of this approach is to search for a suitable derivative asset as the numeraire such that the denominated price processes of primitive assets are martingales under the historical probability measure. It turns out that this numeraire must be the wealth process of the growth optimal portfolio (see Yan, Zhang and Zhang (1999)).

(2) The minimal martingale measure method   This method was introduced in Föllmer-Schweizer (1991) and developed in Schweizer (1995). Assume that the discounted price process of a risky asset is a continuous semimartingale with decomposition $X_t = M_t + A_t$ with M being a square-integrable martingale and A being a process with square-integrable variation. If there exists a martingale measure P^* such that for any square-integrable P-martingale which is orthogonale to M under P remains a martingale under P^*, then we call P^* the minimal martingale measure. We define the price at 0 of a contingent claim ξ by $x^* = E^*[\widetilde{\xi}]$. Then

$$\widetilde{\xi} = x^* + \int_0^T \pi_s^* dX_s + L_T,$$

where L is a square-integrable P-martingale which is orthogonal to M, and (x^*, π^*) attains the following minimum

$$\inf_{x,\pi} E[(\widetilde{\xi} - x - \int_0^T \pi_s dX_s)^2].$$

(3) Esscher transform method The Esscher transform approach is due to Gerber and Shiu (1994) and further developed by Bühlmann et al. (1996). In this approach one assumes that the stock price at time t is $S_t = S_0 e^{X_t}$, where X_t is a Lévy process. Then for any $h \in R$,

$$E[e^{hX_t}] = E[e^{hX_1}]^t \hat{=} e^{t\psi(h)},$$

and $M_t^h = \frac{e^{hX_t}}{e^{t\psi(h)}}$ is a martingale with $M_0^h = 1$. So we can define an equivalent probability measure P^h by $\frac{dP^h}{dP}\Big|_{\mathcal{F}_t} = M_t^h$. Since the discounted price of the stock is $\widetilde{S}_t = e^{-rt}S_t = S_0 e^{X_t - rt}$, P^{h^*} is an equivalent martingale measure iff

$$\widetilde{S}_t M_t^{h^*} = S_0 \frac{e^{(1+h^*)X_t - rt}}{e^{t\psi(h^*)}}$$

is a P-martingale, i.e. h^* satisfies the following equation:

$$\psi(h^*) = \psi(1 + h^*) - r.$$

(4) The utility maximization method Davis (1997) proposed a utility maximization method, in which a "marginal rate of substitution" argument is used to determine a fair price of the contingent claim. Let $\mathcal{T}$ denote the set of self-financing trading strategy. For $\pi \in \mathcal{T}$, $X_x^\pi(T)$ denote the wealth at time T of the trading strategy π with initial welth x, and set

$$V(x) = \sup_{\pi \in \mathcal{T}} E[U(X_x^\pi(T)],$$

where U is a utilitly function. For $\delta, p > 0$, put

$$W(\delta, p, x) = \sup_{\pi \in \mathcal{T}} E\left[X_{x-\delta}^\pi + \frac{\delta}{p}\xi\right].$$

If $\hat{\pi}(x)$ is a unique solution of the equation

$$\frac{\partial W}{\partial \delta}(0, p, x) = 0,$$

then $\hat{\pi}(x)$ is called the fair price of ξ at time 0. It was proved in Davis (1997) that if $V'(x) > 0$ on R_+, then

$$\hat{\pi}(x) = \frac{E[U'(X_x^{\pi^*}(T)\xi]}{V'(x)},$$

where π^* is the optimal strategy which attains $V(x)$. It is easy to see that the Davis's approach coincides with the numeraire portfolio approach when the utility function is the logarithm function.

Recently, Xia and Yan (2000) showed that this fair price $\hat{\pi}(x)$ can be represented by $\mathbf{E_Q}[\widetilde{\xi}]$ with a martingale measure $\mathbf{Q}$ and has an economic meaning that if the price of ξ in the market is p We will show that the investor should buy certain shares of contingent claim B at the price p.

ACKNOWLEDGMENT.   This note is based on an invited lecture presented in IMS Workshop on Applied Probability, organized by the Institute of Mathematical Sciences, the Chinese University of Hong Kong, held on May 31–June 12, 1999. The author wishes to thank the organizing committee for the kind invitation.

References

[1] nsel, J.-P. and C. Stricker (1993) Unicité et existence de la loi minimale. Sém. Probab. XXVII, LN in Math. **1557**, Springer, 22-29.

[2] ajieux, I. and R. Portait (1995a) The numeraire portfolio: a new approach to continuous time finance. Working paper, August 1995, George Washington Univ. and E.S.S.E.C.

[3] ajieux, I. and R. Portait (1995b) Pricing contingent claims in incomplete markets using the numeraire portfolio. Working paper, November 1995, George Washington Univ. and E.S.S.E.C.

[4] lack, F. and M. Scholes (1973) The Pricing of Options and Corporate Liabilities. Journal of Political Economy **3**, 637-654.

[5] ühlmann, H., F. Delbaen, P. Embrechts and A. N. Shiryaeve (1996) No-arbitrage, change of measure and conditional Esscher transforms. CWI Quarterly **9**(4),291-317.

[6] avis, H. A. M. (1997) Option pricing in incomplete markets, in: Mathematics of Derivative Securities. Dempster, A. H. and Pliska, S. R. (eds.), Publications of the Newton Institute, Cambridge Univ. Press, 216-226.

[7] elbaen, F. and W. Schachermayer (1994) A general version of the fundamental theorem of asset pricing. Math. Ann. **300**, 463-520.

[8] öllmer, H., and M. Schweizer (1991) Hedging of contingent claims under incomplete information. Appl. Stoch. Analysis, M.H.A. Davis and R.J. Elliott (eds.), Stoch. Monographs, Vol. 5, Gordon & Breach., 389-414.

[9] arman, M., and S.W. Kohlhagen (1983) Foreign currency option values. J. Intern. Money and Finance **2**, 231-237.

[10] eman, M., N. El Karoui and J. Rochet (1995) Change of numeraire, change of probability measure and option pricing. J. Appl. Probab. **32**, 443-458.

[11] erber, H. U. and E. S. W. Shiu (1994) Option pricing by Esscher transform. Trans. Soc. Actuaries **46**, 51-92.

[12] oldman, M. B., H. B. Sosin and M. A. Gatto (1979) Path dependent options: buy at low, sell at the high. J. Finance **34**, 1111-1128.

[13] arrison, J. M. and D. Kreps (1979) Martingales and Arbitrage in Multiperiod Securities Markets. Journal of Economic Theory **20**, 381-408.

[14] arrison, J. M. and S. Pliska (1981) Martingales and Stochastic Integrals in the Theory of Continuous Trading. Stoch. Proc. and their Appl. **11**, 215-260.

[15] acka, S. D. (1992) A martingale representation result and an application to incomplete financial market. Math. Finance **2**, No. 4, 239-250.

[16] aratzas, I.(1996) Lectures on the Mathematics of Finance. CRM Monographs **8**, American Mathematical Society.

[17] ong, J. B. (1990) The numeraire portfolio. J. Fin. Econ. **26**(1), 29-69.

[18] usiela, M. and M. Rutkowski (1997) Martingale Methods in Financial Modeling. Springer-Verlag.

[19] chweizer, M. (1995) On the minimal martingale measure and the Föllmer- Schweizer decomposition. Stochastic Analysis and Applications **13**, 573-599.

[20] hiryaev, A. N. (1999) Essentials of Stochastic Finance: Facts, Models, Theory. World Scientific.

[21] tricker, C. and J. A. Yan (1998) Some remarks on the optional decomposition theorem. Séminaire de Probab. XXXII, LN in Math. 1686, Springer, 56-66.

[22] ia, J. M. and J. A. Yan (2000) Martingale Measure Method for Expected Utility Maximization and Valuation in Incomplete Markets. Preprint.

[23] an, J. A. (1998a) Introduction to Martingale Methods in Option Pricing. LN in Math. **4**, Liu Bie Ju Centre for Math. Sciences, City University of Hong Kong.

[24] an, J. A. (1998b) A new look at the fundamental theorem of asset pricing. J. Korean Math. Soc. **35**, 659-673.

[25] an, J. A., S. G. Zhang and Q. Zhang (1999) Growth optimal portfolio in a market driven by a jump-diffusion-like process or a Lévy process. Preprint.

INSTITUTE OF APPLIED MATHEMATICS, ACADEMY OF MATHEMATICS AND SYSTEM SCIENCES, ACADEMIA SINICA, BEIJING, 100080, P. R. CHINA, &, LIU BIE JU CENTRE FOR MATHEMATICAL SCIENCES, CITY UNIVERSITY OF HONG KONG, 83 TAT CHEE AVENUE, KOWLOON, HONG KONG

AMS/IP Studies in Advanced Mathematics
Volume 26, 2002

On Comparison Theorems for Diffusion Processes

Xinsheng Zhang

ABSTRACT. It is well known that when two diffusion processes in R^1 have same diffusion terms, the comparison theorems for their sample path can be obtained. But, when two diffusion processes have different diffusion terms, the comparison results on their sample path will not hold in general, only in some special cases. In this paper, We will discuss the problems of comparison between their distributions and their convex order. Several comparison theorems are given for two uniformly elliptic diffusion processes and for two uniformly elliptic diffusion processes with jumps. We also discuss the comparison problems for one kind of infinite dimensional diffusion, named superdiffusion.

1. Introduction

The comparison theorems for diffusion processes is a interesting subject. Many authors have devoted their interest to this area. For example, Anderson[1], Ikeda and Watanabe[14], Skorokhod[19], etc.

When two diffusion processes in R^1 have the same diffusion coefficients, we can obtain the following well known comparison theorem.

Theorem 1.1.[14] Suppose that we are given the following:
(i) a real continuous function $\sigma(t, x)$ defined on $[0, \infty) \times R^1$ such that

$$|\sigma(t, x) - \sigma(t, y)| \leq \rho(|x - y|), \quad x, y \in R^1, \ t > 0,$$

where ρ is a strictly increasing function defined on $[0, \infty)$ such that $\rho(0) = 0$ and

$$\int_{0+} \rho(x)\, dx = \infty,$$

(ii) two real continuous functions $b_1(t, x)$ and $b_2(t, x)$ defined on $[0, \infty) \times R^1$ such that

$$b_1(t, x) \leq b_2(t, x) \quad t \geq 0, x \in R^1.$$

Suppose that $X_i(t)$, $i = 1, 2$ are the (pathwise unique) solutions to the stochastic differential equations

$$dX_i(t) = \sigma(t, X_i(t))dB(t) + b_i(t, X_i(t)); \quad X_i(0) = x_i,$$

Key words and phrases. Comparison Theorems, Uniformly Elliptic Diffusion Processes, Uniformly Elliptic Diffusion Processes with Jumps, Superprocesses.

where $B(t)$ is a one-dimensional Brownian motion with $B(0) = 0$. If $X_1(0) \leq X_2(0)$, then

$$X_1(t) \leq X_2(t) \quad \text{for all t, P a.s..}$$

When two one dimensional diffusion processes have the different diffusion term, O'Brien[17] first proved the same type comparison theorem as following.

Theorem 1.2[17] Let $B(t)$ be a one-dimensional Brownian motion with $B(0) = 0$ and $\sigma_i : R^1 \to (0, \infty)$, $i = 1, 2$ be continuously differentiable functions. Let z^i, $i = 1, 2$, be constants. Suppose that $X_i(t)$, $i = 1, 2$, be the global solutions of stochastic differential equations

$$dX_i(t) = \sigma_i(X_i(t))dB(t) + \frac{1}{2}\sigma_i(X_i(t))\sigma_i'(X_i(t))dt,$$

with initial data $X_i(0) = z^i$. If

$$\int_{z^1}^{y} \frac{dx}{\sigma_1(x)} \geq \int_{z^2}^{y} \frac{dx}{\sigma_2(x)}, \quad \text{for all } y \in R^1.$$

Then $X_1(t) \leq X_2(t)$ a.s. for all $t \geq 0$.

Obviously, the drift term have deeply dependent on the diffusion term in O'Brien's results. In general, for two d-dimensional diffusion processes with different diffusion term, we cannot get comparison results on their sample path. Therefor, we need to look for other type of comparison theorem. We will consider the comparison of the distributions of two diffusion processes. So, for two d-dimensional diffusion processes $X(t)$, $Y(t)$, we put following comparison problem: Let $\mathcal{F}$ be a sort of function in R^d. Find some sufficient or necessary conditions, such that

$$Ef(X(t)) \leq Ef(Y(t)), \quad \forall f \in \mathcal{F}.$$

Borkar[3] first obtain such type of results for two diffusion processes in R^d with different diffusion coefficients and without drift.

Theorem 1.3[3] Let $X(t)$ be a d-dimensional diffusion process described by the stochastic differential equation

$$X(t) = x + \int_0^t \sigma(X(s)) \, dB(t), \quad x \in R^d, \ t \geq 0$$

where $B(t)$ is a d-dimensional standard Brownian motion and $\sigma : R^d \to R^{d \times d}$ is bounded continuous and satisfies

$$\lambda^2 \|y\|^2 \leq \|\sigma(x)^T y\|^2 \leq \Lambda^2 \|y\|^2, \quad x, y \in R^d,$$

for some $0 < \lambda < \Lambda$. Let

$$G = \{f : R^d \to R | f \text{ is convex function and satisfies (1.1) below}\},$$

$$\lim_{\|x\| \to \infty} e^{-\alpha \|x\|^2} |f(x)| = 0 \quad \text{for all } \alpha > 0. \tag{1.1}$$

Then for any $T > 0$ and $f_1, f_2 \in G$

$$\begin{aligned}
& E[\int_0^T f_1(x + \lambda B(t))dt + f_2(x + \lambda B(T))] \\
\leq \ & E[\int_0^T f_1(X(t)dt + f_2(X(T))] \\
\leq \ & E[\int_0^T f_1(x + \Lambda B(t))dt + f_2(x + \Lambda B(T))]
\end{aligned}$$

Recently, Chen and Wang[4] discussed the problem order-preservation and positive correlation for multidimensional diffusion processes. They obtained the comparison results for two diffusion semigroups for all monotone functions.

Let

$$A_l = \frac{1}{2} \sum_{i,j}^{d} a_{i,j}^{(l)} \frac{\partial^2}{\partial x_i \partial x_j} + \sum_{i=1}^{d} b_i^{(l)} \frac{\partial}{\partial x_i}, \quad l = 1, 2.$$

and let $\{P_t^{(l)}\}$ be the semigroup generated by $A^{(l)}$. We call $P_t^{(1)} \geq P_t^{(2)}$, if for all bounded continuous monotone functions f, and all $x \geq y$ and $t \geq 0$,

$$P_t^{(1)} f(x) \geq P_t^{(2)} f(y)$$

Theorem 1.4[4] $P_t^{(1)} \geq P_t^{(2)}$ if and only if the following two conditions hold:
(i) for all i and j, $a_{i,j}^{(1)} = a_{i,j}^{(2)}$ and $a_{i,j}^{(1)}(x)$ depends only on x_i and x_j.
(ii) for all i, $b_i^{(1)}(x) \geq b_i^{(2)}(y)$ whenever $x \geq y$ with $x_i = y_i$.

In this paper, first, we will give some new comparison theorem for the d-dimensional diffusion $X(t)$. Then, we will extend the Borkar's result to uniformly elliptic diffusion with jumps. Finally, Two comparison theorems are given for one kind of infinite dimensional diffusion, named superdiffusion.

The main tool we used is following abstract integral by part formula:

Lemma 1[21] Let T_t, S_t be strongly continuous semigroups on the Banach space E, with infinitesimal generators A and B, respectively. Suppose that $x \in \mathcal{D}(A)$. If $T_s x \in \mathcal{D}(B)$ for $0 < s < t$ and $\int_0^t \|BT_s x\| \, ds < \infty$, then we have

$$S_t x - T_t x = \int_0^t S_{t-s}(B - A) T_s x \, ds$$

Throughout this paper, we will suppose that $\sigma : R^d \to R^{d \times d}$ be bounded continuous matrices value functions. Let $a(x) = \sigma(x)\sigma^T(x)$, $a(x) = (a_{i,j}(x))_{d \times d}$ satisfies the following

Condition (A):

$$\lambda^2 \|y\|^2 \leq y^T a(x) y \leq \Lambda^2 \|y\|^2, \quad x, y \in R^d, \tag{1.2}$$

for some $0 < \lambda < \Lambda$, and $a_{i,j}(x)$, $1 \leq i \leq d$, $1 \leq j \leq d$ are bounded continuous functions on R^d.

Let $B(t)$ be a d-dimensional standard Brownian motion. We call $X(t)$ is uniformly elliptic diffusion in R^d, if $X(t)$ is the solution of the following stochastic differential equation:

$$dX(t) = \sigma(X(t))dB(t) + b(X(t))dt.$$

where $\sigma(x)$ satisfies condition (A).

2. Comparison Theorems for Uniformly Elliptic Diffusion

Let $X(t)$ be d-dimensional diffusion processes without drift, i.e., $X(t)$ satisfies the following stochastic differential equation:

$$dX(t) = \sigma(X(t))\, dB(t), \quad \text{with } X(0) = x. \tag{2.1}$$

We have the following comparison theorem for $X(t)$:

Theorem 2.1[23] Let $X(t)$ be a d-dimensional uniformly elliptic diffusion i.e., σ satisfies (1.2). If $d \geq 2$, then $\forall t > 0$, $\forall C > 0$, $\forall x \in R^d$

$$P_x(|X(t)| \geq C) \leq 2P_x(|B(t)| \geq C/\Lambda), \tag{2.2}$$

$$P_x(\sup_{0\leq t\leq T} |X(t)| \geq C) \leq 2P_x(\sup_{0\leq t\leq T} |B(t)| \geq C/\Lambda), \quad \forall T > 0. \tag{2.3}$$

Sketch of Proof: Set $R(t) = |B(t)|$, and $\bar{R}(t) = |X(t)|$. By applying Itô formula, we can express $R(t)$ and $\bar{R}(t)$ as following

$$R(t) = |x| + \int_0^t d\beta(u) + \frac{1}{2}\int_0^t (d-1)R(u)^{-1}\, du,$$

where $\beta(u)$ is a one-dimensional standard Brownian motion.

$$\bar{R}(t) = |x| + \int_0^t H(u)\, dW(u) + \int_0^t h(X(u))\, du,$$

where

$$H(u)^2 = \bar{R}(u)^{-2} X(u)^T \sigma(X(u))\sigma^T(X(u))X(u),$$

$$h(u) = \frac{[Trace\, \sigma(X(u))\sigma^T(X(u)) - \bar{R}(u)^{-2}X(u)^T\sigma(X(u))\sigma^T(X(u))X(u)]}{2\bar{R}(u)}.$$

For $\forall \varepsilon > 0$, set

$$b(x) = \begin{cases} (d-1)/(2y), & y \geq \varepsilon \\ (d-1)y/(2\varepsilon^2), & y \in (0, \varepsilon) \\ 0, & y \leq 0. \end{cases}$$

We suppose that $\Lambda = 1$ and consider the following stochastic differential equation:

$$Z(t)^\varepsilon = |x| + \int_0^t H(u)\, dW(u) + \int_0^t b(Z(u)^\varepsilon)\, du.$$

Let $\bar{\tau}_\varepsilon = inf\{t > 0,\ \bar{R}(t) \leq \varepsilon\}$, $\tau_\varepsilon = \{t > 0,\ Z^\varepsilon(t) \leq \varepsilon\}$, $Y(t) = \bar{R}(t \wedge \tau_\varepsilon \wedge \bar{\tau}_\varepsilon) - Z^\varepsilon(t \wedge \tau_\varepsilon \wedge \bar{\tau}_\varepsilon)$. By applying Tanaka formula we can proof

$$P_x(Z^\varepsilon(s) \geq \bar{R}(s),\ s \leq \tau_\varepsilon \wedge \bar{\tau}_\varepsilon) = 1$$

Now by using the idea of Hajek[12], for $i = 1, 2$, we consider the following stochastic differential equation:

$$Y_i^\varepsilon(t) = |x| + \int_0^t b(Y_i^\varepsilon(u))du + \int_0^t H(u)dW(u) + (-1)^i \int_0^t (1 - H^2(u))^{\frac{1}{2}}dB(u),$$

where $B(u)$ is a one-dimensional Brownian motion and independent of $W(u)$. Let $\bar{Y}^\varepsilon(t) = \frac{1}{2}[Y_1^\varepsilon(t) + Y_2^\varepsilon(t)]$, $T_i^\varepsilon = \inf\{t > 0,\ Y_i^\varepsilon(t) \leq \varepsilon\}$, $i = 1, 2$. We can proof:

$$P_x(Z^\varepsilon(s) \leq \bar{Y}^\varepsilon(s),\ s \leq T_1^\varepsilon \wedge T_2^\varepsilon \wedge \tau_\varepsilon) = 1,$$

and for fixed $t > 0$, $\forall \delta > 0$, $\exists \varepsilon > 0$ such that

$$P_x(T_1^\varepsilon \wedge T_2^\varepsilon \wedge \tau_\varepsilon \leq t) \leq \delta, \quad x \in R^d, \, x \neq 0.$$

So, for $\forall C > 0$, we have

$$
\begin{aligned}
& P_x(\bar{R}(t) \geq C) \\
\leq \;& P_x(\bar{R}(t) \geq C, \tau_\varepsilon \wedge \bar{\tau}_\varepsilon > t) + \delta \\
\leq \;& P_x(Z^\varepsilon(t) \geq C, \tau_\varepsilon \wedge \bar{\tau}_\varepsilon > t) + \delta \\
\leq \;& P_x(Z^\varepsilon(t) \geq C, T_1^\varepsilon \wedge T_2^\varepsilon \wedge \tau_\varepsilon > t) + 2\delta \\
\leq \;& P_x(max(Y_1^\varepsilon(t), Y_2^\varepsilon(t)) \geq C, T_1^\varepsilon \wedge T_2^\varepsilon \wedge \tau_\varepsilon > t) + 2\delta \\
\leq \;& P_x(Y_1^\varepsilon(t) \geq C, T_1^\varepsilon \wedge T_2^\varepsilon \wedge \tau_\varepsilon > t) \\
+ \;& P_x(Y_2^\varepsilon(t) \geq C, T_1^\varepsilon \wedge T_2^\varepsilon \wedge \tau_\varepsilon > t) + 2\delta \\
\leq \;& 2P_x(R(t) \geq C, T_1^\varepsilon \wedge T_2^\varepsilon \wedge \tau_\varepsilon > t) + 2\delta.
\end{aligned}
$$

Let $\delta \to 0$, we get

$$P_x(|X(t)| \geq C) \leq 2P_x(|B(t)| \geq C), \quad \forall C > 0, \quad |x| > 0$$

Similarly, we can proof that

$$P_x(\sup_{0 \leq t \leq T} |X(t)| \geq C) \leq 2P_x(\sup_{0 \leq t \leq T} |B(t)| \geq C/\Lambda), \quad \forall T > 0.$$

As one application of this comparison theorem, we can get the following up-bound estimation of packing measure for the image of $X(t)$.

Theorem 2.2[23] Let $X(t)$ be a d-dimensional uniformly elliptic diffusion $X(t)$ satisfies the conditions of theorem 2.1. If $d \geq 3$, then we have the following upbound estimation of packing measure for the image of $X(t)$.

$$\psi - p(ImX[0,1]) \leq C < \infty, \quad P \text{ a.s.},$$

where $\psi(t) = \frac{t^2}{\log\log(1/t)}$, $ImX[0,1] = \{X(s) : s \in [0,1]\}$.

In order to state our next two comparison theorems for more general diffusion processes, we need the following comparison lemma.

Comparison Lemma 2.3[15] Let

$$L_t^{(i)} = \sum_{i,j}^d a_{ij}^{(i)}(t,x)\frac{\partial^2}{\partial x_i \partial x_j} + \sum_{i=1}^d b_i^{(i)}(t,x)\frac{\partial}{\partial x_i} \quad i = 1,2$$

be the generator of diffusion process $X_t^{(i)}$, and $T_i^{s,t}f(x) = E_{s,x}f(X^{(i)})$, $s \leq t$. Here the coefficients a_{ij}, b_i are bounded measurable functions on $R_+ \times R^d$, moreover, for all $T, C > 0$ the following holds:

$$\inf_{0 \leq s \leq T} \inf_{|x| < C} \inf_{0 \neq z \in R^d} ([a_{ij}(s,x)]z, z)/|z|^2 > 0$$

and

$$\lim_{\delta \to 0} \sup_{0 \leq s \leq T} \sup_{|x|,|y| < C; |x-y| < \delta} ||[a_{ij}(s,x)]_{i,j} - [a_{ij}(s,y)]_{i,j}|| = 0$$

$$\lim_{\delta \to 0} \sup_{0 \leq s \leq T} \sup_{|x|,|y| < C; |x-y| < \delta} |[b_i(s,x)]_i - [b_i(s,y)]_i| = 0$$

Suppose that $f \in C_b(R^d)$ and let s, t with $s < t$ be given such that $T_2^{u,t} f \in C_b^2(R^d)$ for each $u \in [s,t)$. Suppose that a continuous Lebesgue integrable function ϕ from $[s,t)$ to R^+ exists such that

$$|L_r^{(2)} T_2^{s,t} f(x)| \le \phi(t) \quad \text{and} \quad |L_u^{(1)} T_2^{s,t} f(x)| \le \phi(t)$$

for all $r, t \in [s,t)$, and $x \in R^d$. Then

$$T_1^{s,t} f(x) - T_2^{s,t} f(x) = \int_s^t T^{s,u} (L_u^{(1)} - L_u^{(2)}) T_2^{s,u} f(x) \, du$$

for all $x \in R^d$.

Now, we consider the diffusion process $X(t)$ in R^d, whose infinitesimal generator is defined by

$$L f(x) = \frac{1}{2} \triangle + \sum_{i=1}^d b_i(x) \frac{\partial f}{\partial x_i}(x), \tag{2.4}$$

where $\triangle = \sum_{i=1}^d \frac{\partial^2 f}{\partial x_i^2}(x)$.

Comparison Theorem 2.4 Let $X(t)$ be a diffusion processes in R^d defined as above with $X(0) = x_0$ and $B(t)$ be the standard Brownian motion in R^d with $B(0) = x_0$. Let $\bar{F}_t(x)$ and $F_t(x)$ be the distribution of $X(t)$ and $B(t)$, respectively. That is, $\bar{F}_t(x) = P_{x_0}(X(t) < x)$, and $F_t(x) = P_{x_0}(B(t) < x)$. If $b(x) \le 0$, and satisfies the following linear increase condition:

$$|b(x)| \le M|x|, \qquad x \in R^d,$$

where M is a positive constant. Then $\bar{F}_t(x) \le F_t(x)$ $\forall t > 0$, $\forall x \in R^d$.

Comparison Theorem 2.5 Let $X(t)$ be a diffusion process in R^d, whose infinitesimal generator is given by (2.4). Suppose that $b_i(x)$ be twice differentiable. Let $W(t)$ be a Brownian motion with linear drift, whose infinitesimal generator is defined by

$$L_1 f(x) = \frac{1}{2} \triangle f(x) + \sum_{i=1}^d [b_i(0) + \sum_{j=1}^d \frac{\partial b_i}{\partial x_j}(0) x_j] \frac{\partial f}{\partial x_i}(x).$$

Let $F_t^1(x)$ and $F_t(x)$ be the one dimensional distribution of $X(t)$ and $W(t)$ respectively. If $b_i(x)$, $i = 1, 2, \cdots, d$ are convex functions in R^d, and, and satisfies the following linear increase condition:

$$|b(x)| \le M|x|, \qquad x \in R^d,$$

where M is a positive constant. Then $F_t^1(x) \ge F_t(x)$.

Remark: The the point of proof of theorem 2.4 and 2.5 is using comparison lemma 2.3.

3. The Comparison Theorems for Uniformly Elliptic Diffusion with Jumps

In this section, we will consider the diffusion process $X(t)$ with jumps in R^d, whose infinitesimal generator is defined by

$$Lf(x) = \frac{1}{2}\sum_{i,j} a_{i,j}(x)\frac{\partial^2 f}{\partial x_i\,\partial x_j}(x) + \int_{R^d}[f(x+y) - f(x) - \sum_{j=1}^d y_j\frac{\partial f}{\partial x_j}1_{\{|y|<1\}}]\nu(dy)$$

(3.1)

where $(a_{i,j}(x))_{d\times d} = a(x)$ is same as in introduction and ν is a Lévy measure which is a measure on R^d with $\nu(\{0\}) = 0$ and $\int_R^d(1\wedge|y|^2)\nu(dy) < \infty$.

Let $X_1(t)$ and $X_2(t)$ be two special Lévy processes in R^d, whose infinitesimal generators are defined by

$$L_1f(x) = \frac{\lambda^2}{2}\sum_{i=1}^d\frac{\partial^2 f}{\partial x_i^2}(x) + \int_{R^d}[f(x+y) - f(x) - \sum_{j=1}^d y_j\frac{\partial f}{\partial x_j}1_{\{|y|<1\}}]\nu(dy),$$

$$L_2f(x) = \frac{\Lambda^2}{2}\sum_{i=1}^d\frac{\partial^2 f}{\partial x_i^2}(x) + \int_{R^d}[f(x+y) - f(x) - \sum_{j=1}^d y_j\frac{\partial f}{\partial x_j}1_{\{|y|<1\}}]\nu(dy),$$

respectively. For $X(t)$, $X_1(t)$ and $X_2(t)$, we have the following comparison theorem.

Comparison Theorem 3.1: Let f be a convex function from R^d to R with at most polynomial growth at infinity. Then

$$E_x f(X_1(t)) \leq E_x f(X(t)) \leq E_x f(X_2(t))\ \ \forall t > 0, \quad \forall x \in R^d \qquad (3.2)$$

Sketch of the proof

Let $T(t)$, $T_1(t)$ and $T_2(t)$ be the transition semigroups of $X(t)$, $X_1(t)$ and $X_2(t)$, respectively. It is well known that the Lévy processes are homogeneous in space, i.e. $T_i(t)$, $i = 1, 2$ commute with an arbitrary shift operator. So, if f is convex function, then $T_i(t)f$ is also a convex function.

By using the following modification: Let $f : R^d \to R^1$ be convex, set $\rho_n(x) = n^d\rho(nx)$, $x \in R^d$, $n \geq 1$ with

$$\rho(x) = \begin{cases} c\exp\left\{-\frac{1}{1-||x||^2}\right\}, & ||x|| < 1 \\ 0, & \text{otherwise} \end{cases}$$

where c is a positive constant such that $\int_{R^d}\rho(x)\,dx = 1$. Define

$$f_n(x) = \int_{R^d} f(y)\rho_n(x-y)\,dy \quad x \in R^d,\ \ n \geq 1.$$

Then it is easy to prove that

(i) $f_n \in C_0^\infty(R^d)$, $n \geq 1$ and $\frac{\partial^2 f_n}{\partial x_i\,\partial x_j} \in C_0^\infty(R^d)$, $1 \leq i, j \leq d$, $n \geq 1$.

(ii) f_n is convex on R^d, $n \geq 1$.

(iii) f_n converges to f uniformly on any compact subset of R^d as $n \to \infty$.

So, we only to prove our result for a convex function f with $f \in C_0^\infty(R^d)$. In this case, one can check that all conditions of Lemma 1 hold. By lemma 1 and the following lemma 3.2[3], we have

$$T(t)f(x) - T_i(t)f(x) = \int_0^t T(s)(L - L_i)T_i(s)f(x)ds \quad i = 1, 2.$$

Note that:

$$(L - L_1)T_1(s)f(x)$$
$$= \sum_{i,j} a_{i,j}(x)\frac{\partial^2 T_1(s)f(x)}{\partial x_i \partial x_j} - \lambda^2 \triangle T_1(s)f(x)$$
$$\geq \lambda^2 \triangle T_1(s)f(x) - \lambda^2 \triangle T_1(s)f(x)$$
$$= 0.$$

$$(L - L_2)T_2(s)f(x)$$
$$= \sum_{i,j} a_{i,j}(x)\frac{\partial^2 T_2(s)f(x)}{\partial x_i \partial x_j} - \Lambda^2 \triangle T_2(s)f(x)$$
$$\leq \Lambda^2 \triangle T_2(s)f(x) - \Lambda^2 \triangle T_2(s)f(x)$$
$$= 0.$$

Therefore:

$$T(t)f(x) \geq T_1(t)f(x), \quad T(t)f(x) \leq T_2(t)f(x).$$

i.e:

$$E_x f(X_1(t)) \leq E_x f(X_t) \leq E_x f(X_2(t)).$$

Lemma 3.2[3] Let $B = (b_{ij})_{d \times d}$ $C = (c_{ij})_{d \times d}$ be non-negative matrices. Moreover, B satisfies the following condition : $\exists 0 < \lambda < \Lambda < +\infty$ for $\forall \xi \in R^d$,

$$\lambda^2 \xi^T \xi \leq \xi^T A \xi \leq \Lambda^2 \xi^T \xi$$

Then

$$\lambda^2 \sum_{i=1}^d b_{ii} \leq \sum_{i,j} a_{ij}b_{ij} \leq \Lambda^2 \sum_{i=1}^d b_{ii}.$$

Using same idea, you also can prove the following results.

Comparison theorem 3.3[22] Let $X(t), Y_1(t), Y_2(t)$ be three diffusion processes in R^d, which are defined by the following stochastic differential equations

$$dX(t) = \sigma(X(t))dW(t) + AX(t)dt, \quad X(0) = x$$
$$dY_1(t) = \lambda^2 dW(t) + AY_1(t)dt, \quad Y_1(0) = x$$
$$dY_2(t) = \Lambda^2 dW(t) + AY_2(t)dt, \quad Y_2(0) = x$$

where A is a $d \times d$ order matrix, $W(t)$ is d-dimensional Brownian motion, and σ satisfies condition (A). Let f be a convex function from R^d to R with at most polynomial growth at infinity. Then

$$E_x f(Y_1(t)) \leq E_x f(X(t)) \leq E_x f(Y_2(t)), \quad \forall x \in R^d.$$

Remark: When $A = 0$, we get the result of Borkar's. But our approach is much different from Borkar's. He use Itô formula as the main tool. It is also not easy to get our result by Borkar's method as I know.

It is similar as in the case of diffusion processes without jumps in section 2, we also can obtained the comparison results on their distributions for two diffusion processes with jumps. Let $X(t)$ be a diffusion processes with jumps in R^d, which is defined by the following stochastic differential equation

$$dX(t) = b(X(t))dt + dZ(t), \quad X(0) = x_0, \tag{3.3}$$

where $Z(t)$ is a d-dimensional Lévy process, and $b(x)$ is a d-dimensional vector valued function.

Comparison Theorem 3.4 Let $X(t)$ be a diffusion processes with jumps in R^d defined as above and $Z(t)$ be the Lévy process in R^d with $Z(0) = x_0$. Let $\bar{F}_t(x)$ and $F_t(x)$ be the distribution of $X(t)$ and $Z(t)$, respectively. That is, $\bar{F}_t(x) = P_{x_0}(X(t) < x)$, and $F_t(x) = P_{x_0}(Z(t) < x)$. If $b(x) \leq 0$, then $\bar{F}_t(x) \leq F_t(x)$ $\forall t > 0, \forall x \in R^d$.

Comparison Theorem 3.5 Let $X(t)$ be a diffusion process with jumps in R^d, defined by (3.3). Suppose that $b_i(x)$ be twice differentiable. Let $Y(t)$ be a Lévy processes with linear drift, which is defined by the following stochastic differential equation

$$dY(t) = \sum_{i=1}^{d}[b_i(0) + \sum_{j=1}^{d} \frac{\partial b_i}{\partial x_j}(0)Y_i(t)]dt + dZ(t). \tag{3.4}$$

Let $F_t^1(x)$ and $F_t(x)$ be the one dimensional distribution of $X(t)$ and $Y(t)$ respectively. If $b_i(x)$, $i = 1, 2, \cdots, d$ are convex functions in R^d, and $X(0) = Y(0)$, then $F_t^1(x) \geq F_t(x)$.

Remark: The proof of comparison theorem 3.3 and comparison theorem 3.4 is similar as the proof of comparison theorem 2.4 and 2.5. The key point is to extend the comparison lemma 2.3 to the case of Lévy process.

Lemma 3.6 Let $X(t)$ be diffusion process with jumps in R^d, which is defined by stochastic differential equation (3.3). Let $T(t)$ and L be its transition semigroup and infinitesimal generator, respectively, $T_1(t)$ and L_1 be the transition semigroup and infinitesimal generator of Lévy process $Z(t)$, respectively. For $f \in C_b(R^d)$, if $T_1(t)f \in C_b^1(R^d)$. Then,

$$T(t)f - T_1(t)f = \int_0^t T_1(s)[L_1 - L]T(s)f(x)\,ds.$$

4. Two Comparison Theorems for Dawson-Watanabe Superprocess

(A) Dawson-Watanabe superprocess

Let $(E, \mathcal{E})$ be a polish space. Let $\mathcal{M}(E)$ denote the set of all finite measure on E, and $bC(E)$ denote (respectively, $bpC(E)$) the family of bounded continuous functions(respectively, bounded positive continuous functions) on E. Set:

$$\Psi(x, \lambda) = -b(x)\lambda - c(x)\lambda^2 + \int_0^\infty (1 - e^{-\lambda u} - \lambda u)n(x, du)$$

where $c \geq 0$, and b are bounded measurable functions and $n(x, du)$ is a kernel from $E \times R_+$ to R_+ such that $\int_0^\infty u \wedge u^2 n(x, du)$ is a bounded function on E. Here we give two special cases of $\Psi(x, \lambda)$:

(1) $\Psi(x, \lambda) = b(x)\lambda - c(x)\lambda^2$, where $c(x) \geq 0$;

(2) $\Psi(x, \lambda) = -\gamma(x)\lambda^{1+\beta}$, where $\gamma(x) \geq 0$, $0 < \beta \leq 1$.

Let ξ_t be Borel right process on E, T_t denote its transition semigroup. By Fitzsimmons's [9] results, there exists a unique $\mathcal{M}(E)$-valued Markov process X_t, whose Laplace functional is determined by the following formula:

$$E_\mu \exp\{- < X_t, f >\} = \exp\{- < \mu, u(t)f >\}, \quad f \in bpC(E), \quad \mu \in \mathcal{M}(E),$$

where $u(t)f$ is the solution of the integral equation:

$$u(t)f = T_t f + \int_0^t T_{t-s}\Psi(u(s)f)ds.$$

Moreover, X_t has the following branching property:

$$(X_t : t \geq 0 | X_0 = \mu + \nu) \overset{d}{=} (X_t' + X_t'' : t \geq 0 | X_0' = \mu, X_0'' = \nu),$$

" $\overset{d}{=}$ " means the random variables of the two sides of equation have the same distribution. Following Dynkin [8], we call X_t a Dawson-Watanabe superprocess with parameter (ξ, Ψ).

Let:

$$D(\mathcal{L}) = \{\phi(< \mu, f_1 >, \cdots, < \mu, f_n >), \quad \phi \in C_0^\infty(R^n), \quad f_1, \cdots, f_n \in \mathcal{D}(L)\},$$

where $\mathcal{D}(L)$ is the domain of operator L, $< \mu, f_i >= \int_E f_i d\mu$, $i = 1, \cdots, n$.

For $\forall F \in D(\mathcal{L})$, set:

$$\begin{aligned} \mathcal{L}F(\mu) &= \int_{R^d} \mu(dx)c(x)F''(\mu, x) + \int_E \mu(dx)[LF'(\mu, x) - b(x)F'(\mu, x)] \\ &+ \int_E \mu(dx)\int_0^\infty n(x, du)[F(\mu + u\delta_x) - F(\mu) - uF'(\mu, x)], \end{aligned}$$

where $F'(\mu, x) = \lim_{\epsilon \to 0} \frac{F(\mu + \epsilon\delta_x) - F(\mu)}{\epsilon}$, δ_x denote the Dirac measure of point x. $F''(\mu, x)$ can be defined similarly.

By Fitgsinmmon's results(see[9]), we know that $\mathcal{L}$ is the generalized infinitesimal operator of X_t, i.e.

$$F(X_t) - F(X_0) - \int_0^t \mathcal{L}F(X_s)ds, \quad t \geq 0$$

is a locally bounded $\mathcal{P}_\mu$-martingale.

(B) Two comparison theorems

Let $\xi = \{\xi_t, t \geq 0\}$ be a Borel right process on E and

$$\Psi_1(x, \lambda) \equiv -b_1(x)\lambda - c_1(x)\lambda^2 + \int_0^\infty (1 - e^{-\lambda u} - \lambda u)n_1(x, du),$$

$$\Psi_2(x, \lambda) \equiv -b_2(x)\lambda - c_2(x)\lambda^2 + \int_0^\infty (1 - e^{-\lambda u} - \lambda u)n_2(x, du).$$

Theorem 1 establish the relationship between two Dawson-Watanabe superprocess with same ξ but different Ψ.

Theorem 4.1:[24] Suppose $X_t^{(i)}$ $i = 1, 2$ be D-W superprocess with parameters (ξ, Ψ_i). T_t and L are the transition semigroup and infinitesimal generator of ξ_t, respectively. If $\mathcal{D}(L)$(the domain of operator L) is dense in $bpC(E)$ and $\Psi_1(x, \lambda) \leq \Psi_2(x, \lambda)$, then

$$E_\mu \exp\{- < f, X_t^{(1)} >\} \geq E_\mu \exp\{- < f, X_t^{(2)} >\}, \quad f \in bpC(E), \mu \in \mathcal{M}(E).$$

Proof:

Let $\mathcal{P}_t^{(i)}$ be the transition semigroup of $X_t^{(i)}$ and $\mathcal{L}_i$ be the infinitesimal generator of $\mathcal{P}_t^{(i)}$. Note that $\mathcal{P}_t^{(i)}$ is the strongly continuous semigroups on Banach space $B(\mathcal{M}(E))$. Here,

$$B(\mathcal{M}(E)) = \{F(\mu): \sup_{\mu \in \mathcal{M}(E)} |F(\mu)| < +\infty\}.$$

For $\forall f \in \mathcal{D}(L)_+$, set $F(\mu) = \exp\{- < \mu, f >\}$. It is easy to check that $F(\mu) \in D(\mathcal{L}_i)$, $i = 1, 2$. Since

$$\mathcal{P}_t^{(i)} F(\mu) = E_\mu \exp\{- < X_t^{(i)}, f >\} = \exp\{- < \mu, u(t)^{(i)} f >\},$$

where

$$u^{(i)}(t)f = T_t f + \int_0^t T_{t-s} \Psi(u^{(i)}(s)f)\, ds, \quad i = 1, 2.$$

Hence

$$\mathcal{P}_t^{(1)} F(\mu) \in D(\mathcal{L}_2) \quad (f \in \mathcal{D}(L)_+ \Longrightarrow u^{(1)}(t)f \in \mathcal{D}(L)).$$

On the other hand,

$$
\begin{aligned}
F'(\mu, x) &= \lim_{\epsilon \downarrow 0} \frac{F(\mu + \epsilon \delta_x) - F(\mu)}{\epsilon} \\
&= \lim_{\epsilon \downarrow 0} \frac{\exp\{- < f, \mu + \delta_x \epsilon >\} - \exp\{- < f, \mu >\}}{\epsilon} \\
&= -e^{- < f, \mu >} f(x)
\end{aligned}
$$

and $F''(\mu, x) = f^2(x) e^{- < f, \mu >}$. Therefore:

$$\mathcal{L}_i F(\mu) = -e^{- < f, \mu >} \int_E \{\Psi_i(f) + Af\} d\mu, \quad i = 1, 2. \tag{4.1}$$

By [9], we know that

$$F(X_t^{(i)}) - F(X_0^{(i)}) - \int_0^t \mathcal{L}_i F(X_s^{(i)})ds, \quad t \geq 0, \quad i = 1, 2 \tag{4.2}$$

is a locally bounded $\mathcal{P}_\mu$-martingale. Take expectation in two side of (4) and applying the martingale property, we have

$$\mathcal{P}_t^{(i)} F(\mu) - F(\mu) - \int_0^t \mathcal{P}_s^{(i)} \mathcal{L}_i F(\mu)ds = 0, \quad i = 1, 2.$$

Therefore,

$$
\begin{aligned}
\|\tfrac{1}{t}(\mathcal{P}_t^{(i)} F - F) - \mathcal{L}_i F\| &= \|\tfrac{1}{t} \int_0^t [\mathcal{P}_s^{(i)} \mathcal{L}_i F - \mathcal{L}_i F]\, ds\| \\
&\leq \tfrac{1}{t} \int_0^t \|\mathcal{P}_s^{(i)} \mathcal{L}_i F - \mathcal{L}_i F\|\, ds.
\end{aligned}
$$

By (4.1), we have

$$
\begin{aligned}
& \mathcal{P}_s^{(i)} \mathcal{L}_i F - \mathcal{L}_i F \\
&= \mathcal{L}_i \mathcal{P}_s^{(i)} F - \mathcal{L}_i F \\
&= -e^{- < u^{(i)}(t)f, \mu >} \int_E \{\Psi_i(u^{(i)}(t)f) + A(u^{(i)}(t)f)\} d\mu \\
&+ e^{- < f, \mu >} \int_E \{\Psi_i(f) + Af\} d\mu,
\end{aligned} \tag{4.3}
$$

where $u^{(i)}(t)f$ is the solution of the following integral equation:

$$u^{(i)}(t)f = T_t f + \int_0^t T_{t-s} \Psi_i(u^{(i)}(s)f)ds \quad i = 1, 2.$$

From the above equalities and $\mathcal{P}_t^{(i)}$ is the strongly continuous semigroup on Banach space $B(\mathcal{M}(E))$, we can conclude that

$$\frac{1}{t}\int_0^t \|\mathcal{P}_s^{(i)}\mathcal{L}_i F - \mathcal{L}_i F\| ds \to 0 \text{ as } t \to 0, \quad i = 1, 2.$$

Also,

$$
\begin{aligned}
&\int_0^t \|\mathcal{L}_2 \mathcal{P}_s^{(1)} F\| ds \\
= &\int_0^t \| - e^{-<u^{(1)}(s)f,\mu>}\int_E \{\Psi_2(u^{(1)}(s)f) + A(u^{(1)}(t)f)\} d\mu\| ds \\
< &\infty.
\end{aligned}
$$

By lemma 1:

$$
\begin{aligned}
&\mathcal{P}_t^{(2)} F(\mu) - \mathcal{P}_t^{(1)} F(\mu) \\
= &\int_0^t \mathcal{P}_{t-s}^{(2)}(\mathcal{L}_2 - \mathcal{L}_1)\mathcal{P}_s^{(1)} F(\mu) ds \\
= &\int_0^t \mathcal{P}_{t-s}^{(2)} e^{-<\mu,u^{(1)}(s)f>}\int_E [\Psi_1(u^{(1)}(s)f) - \Psi_2(u^{(1)}(s)f)] d\mu\, ds \\
\leq &\, 0,
\end{aligned}
$$

i.e:

$$\exp\{- <\mu, X_t^{(2)}>\} \leq \exp\{- <\mu, X_t^{(1)}>\}.$$

Note that $\mathcal{D}(L)_+$ is dense in $bpC(E)$, we can obtain our result.

When the two Dawson-Watanabe superprocess have the same Ψ and different ξ, we have the following comparison theorem.

Theorem 4.2:[24] Suppose $X_t^{(i)}\ i = 1, 2$ be D-W superprocess with parameters $(\xi^{(i)}, \Psi)$. i.e.

$$E_\mu \exp\{- <f, X_t^{(i)}>\} = \exp\{- <u^{(i)}(t)f, \mu>\}, \quad f \in bpC(E), \quad \mu \in \mathcal{M}(E),$$

where

$$u^{(i)}(t)f = T_t^{(i)}f + \int_0^t T_{t-s}\Psi(u^{(i)}(s)f) ds, \quad i = 1, 2.$$

$T_t^{(i)}$ and L_i are the transition semigroup and infinitesimal generator of $\xi_t^{(i)}$, respectively. If $f \in C_0^\infty(E)$ and $L_1 u^{(1)}(t)f \geq L_2 u^{(1)}(t)f$ (or $L_1 u^{(2)}(t)f \geq L_2 u^{(2)}(t)f$), then

$$E_\mu \exp\{- <f, X_t^{(1)}>\} \leq E_\mu \exp\{- <f, X_t^{(2)}>\}, \quad f \in bpC(E), \quad \mu \in \mathcal{M}(E).$$

Proof:

The theorem 4.2 can be proved by the same way as in the proof of theorem 4.1. First, we need to substitute (4.1) by

$$\mathcal{L}_i F(\mu) = -e^{-<\mu,f>}\int_{R^d} \{\Psi(f) + L_i f\} d\mu, \quad i = 1, 2. \tag{4.4}$$

By (4.4), we have

$$
\begin{aligned}
&\mathcal{P}_s^{(i)}\mathcal{L}_i F - \mathcal{L}_i F \\
= &\, \mathcal{L}_i \mathcal{P}_s^{(i)} F - \mathcal{L}_i F \\
= &\, -e^{-<u^{(i)}(t)f,\mu>}\int_E \{\Psi(u^{(i)}(t)f) + L_i(u^{(i)}(t)f)\} d\mu \\
&\, + e^{-<f,\mu>}\int_E \{\Psi(f) + L_i f\} d\mu,
\end{aligned}
$$

where $u^{(i)}(t)f$ is the solution of the following integral equation:

$$u^{(i)}(t)f = T_i(t)f + \int_0^t T_i(t-s)\Psi(u^{(i)}(s)f)ds \quad i = 1, 2.$$

From the above equalities and $\mathcal{P}_t^{(i)}$ is the strongly continuous semigroup on Banach space $B(\mathcal{M}(E))$, we can conclude that

$$\frac{1}{t}\int_0^t ||\mathcal{P}_s^{(i)}\mathcal{L}_i F - \mathcal{L}_i F|| \, ds \to 0 \text{ as } t \to 0, \quad i = 1, 2.$$

Also,

$$
\begin{aligned}
&\int_0^t ||\mathcal{L}_2 \mathcal{P}_s^{(1)} F|| ds \\
= &\int_0^t || - e^{-<u^{(1)}(s)f,\mu>} \int_E \{\Psi(u^{(1)}(s)f) + L_2(u^{(1)}(t)f)\}d\mu || ds \\
< &\infty.
\end{aligned}
$$

By lemma 1:

$$
\begin{aligned}
&\mathcal{P}_t^{(2)} F(\mu) - \mathcal{P}_t^{(1)} F(\mu) \\
= &\int_0^t \mathcal{P}_{t-s}^{(2)}[\mathcal{L}_2 - \mathcal{L}_1]\mathcal{P}_s^{(1)} F(\mu) ds \\
= &\int_0^t \mathcal{P}_{t-s}^{(2)} e^{-<\mu, u^{(1)}(s)f>} \int_E [L_1(u^{(1)}(s)f) - L_2(u^{(1)}(s)f)]d\mu \, ds \\
\geq &0,
\end{aligned}
$$

i.e:

$$\exp\{-<\mu, X_t^{(2)}>\} \geq \exp\{-<\mu, X_t^{(1)}>\}.$$

Note that $\mathcal{D}(L)_+$ is dense in $bpC(E)$, we can obtain our result.

References

[1]   . J. Anderson, Local behavior of solutions of stochastic integral equations, Transactions A. M. S. 164(1972), 309-321.

[2]   . N. Bhattacharya, , Criteria for recurrence and existence of invariant measures for multidimensional diffusion. Annals of Probability 6 (1978), 541–553.

[3]   . S. Borkar, On comparison theorem for diffusion without drift. Stochastic Processes and Appl. 27 (1987), 245–248.

[4]   . F. Chen and F. Y. Wang, On order-preservation and positive correlation for multidimensional diffusion processes. Probab. Theory Relat. Fields 95(1993), 421-428.

[5]   , A. Dawson, I. Iscoe, and E. A. Perkins, super-Brownian motion: Path Properties and Hitting Probabilities, Prob. Th. Rel. Field, 83(1989), 135-206.

[6]   . A. Dawson, and K. J. Hochberg, The carrying dimension of a stochastic measure diffusion, Ann. Prob., 7(1979),693-703.

[7]   . A. Dawson, The critical measure diffusion, Z. Wahr. verw. Geb. 40(1977), 125-145.

[8]   . B. Dynkin, Three classes of infinite dimensional diffusions, J. Funct. Anal. 86(1989), 75-110.

[9]   . J. Fitzsimmons,   Construction and regularity of measure-valued Markov branching processes, Israel J. Mach., VOL. 64(1988), 337-361.

[10] . I. Gal'čuk, M.H.A. Davis, A note on a comparison theorem for equations with different diffusions. Stochastic 6(1981), 147–149.

[11] . Geiss, and R. Manthey, Comparison results for stochastic differential equations. In: Stochastic processes and optimal control. Engelbert, Karatzas, Röckner eds., Gordon & Breach.

[12] . Hajek, Mean stochastic comparison of diffusion. Z. Wahrsheinlichkeitstheorie verw. Gebiete 68(1985), 315-329.

[13] . Ikeda and S. Watanabe, A comparison theorem for solutions of stochastic differential equations and its applications, Osaka J. Math. 14(1977) 619-633.

[14] . Ikeda and S. Watanabe, Stochastic Differential Equation and Diffusion Processes, Second Edition, North-Holland Publishing Company, Amsterdam · Oxford · New York, 1989.

[15] . Kröger, Comparison Theorems for Diffusion Processes, J. of Theoretical Probability, VOL. 3, No.4(1990), 515-531.

[16] . Y. Lee, T. Y. and W. M. Ni, Global existence, large time behavior and life span of solutions of a semilinear parabolic Cauchy problem, Trans. Amer. Soc. VOL. 333(1992), 365-378.

[17] . L. O'Brien, A new comparison theorem for solution of stochastic differential equations, Stochastic 3(1980), 245-249.

[18] . Pazy, Semigroups of linear operators and applications to partial differential equations, Springer-Verlag, New York, Inc. 1983.

[19] . V. Skorokhod, Studies in the Theory of Random Processes, Addison-Wesley, Reading, Mass-chustter, 1965.

[20] . Yamada, On comparison theorem for solutions of stochastic differential equations and its applications, J. Math. Kyoto Univ. 13(1973), 497-512.

[21] . A. Yan, A perturbation theorem for semigroups of linear operators. Lecture Notes in Math., VOL. 1321, Berlin, 89-91, 1987.

[22] . S. Zhang, Existence and uniqueness of invariant probability measure for uniformly elliptic diffusion, Dirichlet Forms and Stochastic Processes, Editors, Z. M. Ma, M. Röckner, J. A. Yan; Walter de Gruyter · Berlin · New York 1995.

[23] . K. Wang, X. S. Zhang, and Z. B. Li, One comparison theorem for uniformly elliptic diffusion processes and applications, Science in China(Series A), VOL. 23(1993), No. 11, 1121-1129.

[24] . S. Zhang, On the comparison Theorems for the Non-linear Evolution Equation, Lecture Notes in Pure and Applied Mathematics, vol.176, 1996.

DEPARTMENT OF STATISTICS, EAST CHINA NORMAL UNIVERSITY

Titles in This Series

26 **Raymond Chan, Yue-Kuen Kwok, David Yao, and Qiang Zhang, Editors,** Applied Probability, 2002

25 **Donggao Deng, Daren Huang, Rong-Qing Jia, Wei Lin, and Jian Zhong Wong, Editors,** Wavelet Analysis and Applications, 2002

24 **Jane Gilman, William W. Menasco, and Xiao-Song Lin, Editors,** Knots, Braids, and Mapping Class Groups—Papers Dedicated to Joan S. Birman, 2001

23 **Cumrun Vafa and S.-T. Yau, Editors,** Winter School on Mirror Symmetry, Vector Bundles and Lagrangian Submanifolds, 2001

22 **Carlos Berenstein, Der-Chen Chang, and Jingzhi Tie,** Laguerre Calculus and Its Applications on the Heisenberg Group, 2001

21 **Jürgen Jost,** Bosonic Strings: A Mathematical Treatment, 2001

20 **Lo Yang and S.-T. Yau, Editors,** First International Congress of Chinese Mathematicians, 2001

19 **So-Chin Chen and Mei-Chi Shaw,** Partial Differential Equations in Several Complex Variables, 2001

18 **Fangyang Zheng,** Complex Differential Geometry, 2000

17 **Lei Guo and Stephen S.-T. Yau, Editors,** Lectures on Systems, Control, and Information, 2000

16 **Rudi Weikard and Gilbert Weinstein, Editors,** Differential Equations and Mathematical Physics, 2000

15 **Ling Hsiao and Zhouping Xin, Editors,** Some Current Topics on Nonlinear Conservation Laws, 2000

14 **Jun-ichi Igusa,** An Introduction to the Theory of Local Zeta Functions, 2000

13 **Vasilios Alexiades and George Siopsis, Editors,** Trends in Mathematical Physics, 1999

12 **Sheng Gong,** The Bieberbach Conjecture, 1999

11 **Shinichi Mochizuki,** Foundations of p-adic Teichmüller Theory, 1999

10 **Duong H. Phong, Luc Vinet, and Shing-Tung Yau, Editors,** Mirror Symmetry III, 1999

9 **Shing-Tung Yau, Editor,** Mirror Symmetry I, 1998

8 **Jürgen Jost, Wilfrid Kendall, Umberto Mosco, Michael Röckner, and Karl-Theodor Sturm,** New Directions in Dirichlet Forms, 1998

7 **D. A. Buell and J. T. Teitelbaum, Editors,** Computational Perspectives on Number Theory, 1998

6 **Harold Levine,** Partial Differential Equations, 1997

5 **Qi-keng Lu, Stephen S.-T. Yau, and Anatoly Libgober, Editors,** Singularities and Complex Geometry, 1997

4 **Vyjayanthi Chari and Ivan B. Penkov, Editors,** Modular Interfaces: Modular Lie Algebras, Quantum Groups, and Lie Superalgebras, 1997

3 **Xia-Xi Ding and Tai-Ping Liu, Editors,** Nonlinear Evolutionary Partial Differential Equations, 1997

2.2 **William H. Kazez, Editor,** Geometric Topology, 1997

2.1 **William H. Kazez, Editor,** Geometric Topology, 1997

1 **B. Greene and S.-T. Yau, Editors,** Mirror Symmetry II, 1997